gRPC Go for Professionals

Implement, test, and deploy production-grade microservices

Clément Jean

BIRMINGHAM—MUMBAI

gRPC Go for Professionals

Group Product Manager: Gebin George

Product Manager: Kunal Sawant

Senior Editor: Kinnari Chohan

Technical Editor: Maran Fernandes

Copy Editor: Safis Editing

Project Coordinator: Manisha Singh

Proofreader: Safis Editing

Indexer: Hemangini Bari

Production Designer: Shyam Sundar Korumilli

Developer Relations Marketing Coordinator: Sonia Chauhan

First published: July 2023

Production reference: 1230623

Published by Packt Publishing Ltd.
Livery Place
35 Livery Street
Birmingham
B3 2PB, UK.

ISBN 978-1-83763-884-0

www.packtpub.com

To my mother, Géraldine Seyte, for building up my sense of curiosity and determination. To my father, Marc Jean, for exemplifying the power of working hard and smart. To my wife, 李梦昕, for accompanying me throughout the journey.

– Clément Jean

Contributors

About the author

Clément Jean is the CTO of Education for Ethiopia, a start-up focusing on educating K-12 students in Ethiopia. On top of that, he is also an online instructor (on Udemy, Linux Foundation, and others) teaching people about different kinds of technologies. In both his occupations, he deals with technologies such as gRPC and how to apply them to real-life use cases. His overall goal is to empower people through education and technology.

I want to thank my family and friends for supporting me. Without you, this book would not have been possible.

About the reviewers

Wisnu Anggoro is a seasoned backend engineer with over 10 years of experience in the development and maintenance of distributed systems. His strong command of the Golang (Go Programming Language) and microservices architecture enables him to tackle intricate software challenges and implement creative solutions. Wisnu's expertise lies in finding innovative approaches to complex problems within the realm of software development.

Vincent Youmans has 15 years of senior software engineering experience, having worked at renowned companies such as IBM, Capital One, VMware, and various others. Currently, he has been dedicated to Golang for the past 5 years, holding AWS Architect and AWS Developer Associate Certifications. Vincent's expertise lies in HealthTech and FinTech, where he approaches projects with a microservices mindset, leveraging messaging, NoSQL databases, and CI/CD practices. He specializes in gRPC and Protocol Buffers, utilizing them extensively.

Fattesingh Rane, a passionate Software Engineer from India since 2019, thrives on exploring new technologies and building captivating projects. His exceptional skills and enthusiasm make a lasting impact, inspiring others to embrace the endless possibilities of technology.

Table of Contents

Preface   xi

1

Networking Primer   1

Prerequisites   1
Understanding HTTP/2   2
RPC operations   3
Send Header   4
Send Message   5
Send Half Close   6
Send Trailer   6

RPC types   7
Unary   7
Server streaming   8

Client streaming   9
Bidirectional streaming   10

The life cycle of an RPC   11
The connection   13
The client side   14
The server side   15

Summary   16
Quiz   17
Answers   18

2

Protobuf Primer   19

Prerequisites   19
Protobuf is an IDL   20
Serialization and deserialization   22
Protobuf versus JSON   23
Serialized data size   23
Readability   24
Schema strictness   25

Encoding details   26

Fixed-size numbers   27
Varints   28
Length-delimited types   31
Field tags and wire types   32

Common types   34
Well-known types   34
Google common types   35

Services   35

Summary	36	Answers	38
Quiz	37		

3

Introduction to gRPC 39

Prerequisites	39	REST	49
A mature technology	40	GraphQL	49
What is gRPC doing?	42	Comparison with gRPC	50
The server	44	Summary	52
The client	46	Quiz	52
The read/write flow	47	Answers	53
Why does gRPC matter?	49		

4

Setting Up a Project 55

Prerequisites	55	Client boilerplate	68
Creating a .proto file definition	56	Bazel	70
Generating Go code	58	Server and Dial options	70
Protoc	58	grpc.Creds	71
Buf	59	grpc.*Interceptor	71
Bazel	60	Summary	72
Server boilerplate	63	Quiz	73
Bazel	66	Answers	73

5

Types of gRPC Endpoints 75

Technical requirements	75	Inspecting the generated code	78
Using the template	75	Registering a service	79
A Unary API	76	Implementing AddTask	83
Code generation	78	Calling AddTask from a client	84

Bazel   86

The server streaming API   86
Evolving the database   87
Implementing ListTasks   88
Calling ListTasks from a client   90

The client streaming API   92
Evolving the database   92
Implementing UpdateTasks   94

Calling UpdateTasks from a client   95

The bidirectional streaming API   98
Evolving the database   99
Implementing DeleteTasks   99
Calling UpdateTasks from the client   101

Summary   103
Quiz   103
Answers   104

6

Designing Effective APIs   105

Technical requirements   105
Choosing the right integer type   105
An alternative to using integers   108

Choosing the right field tag   109
Required/optional   109
Splitting messages   111
Improving UpdateTasksRequest   113

Adopting FieldMasks to reduce
the payload   115

Improving ListTasksRequest   116

Beware the unpacked repeated field   119
Packed repeated fields   119
Unpacked repeated fields   119

Summary   121
Quiz   121
Answers   122

7

Out-of-the-Box Features   123

Technical requirements   123
Handling errors   124
Bazel   128
Canceling a call   129
Specifying deadlines   133
Sending metadata   134
External logic with interceptors   137
Compressing the payload   142

Securing connections   146
Bazel   148

Distributing requests with load
balancing   149
Summary   153
Quiz   154
Answers   154
Challenges   155

8

More Essential Features   157

Technical requirements	157	Bazel	179
Validating requests	157	**Securing APIs with rate limiting**	**179**
Buf	161	Bazel	182
Bazel	162	**Retrying calls**	**182**
Middleware = interceptor	**164**	Bazel	184
Authenticating requests	**165**		
Bazel	168	**Summary**	**185**
		Quiz	**185**
Logging API calls	**168**	**Answers**	**186**
Tracing API calls	**172**	**Challenges**	**186**

9

Production-Grade APIs   187

Technical requirements	187	**Deploying**	**217**
Testing	187	Docker	218
Unit testing	188	Kubernetes	222
Load testing	200	Envoy proxy	224
Debugging	**204**	**Summary**	**230**
Server reflection	204	**Quiz**	**231**
Using Wireshark	209	**Answers**	**231**
Turning gRPC logs on	216	**Challenges**	**231**

Epilogue   232

Index   233

Other Books You May Enjoy   240

Preface

In the highly interconnected world of microservices, gRPC has become an important communication technology. By standing on the shoulders of Protobuf and implementing must-have features during communication, gRPC provides reliable, efficient, and user-friendly APIs. In this book, you will explore why this is the case, how to write these APIs, and how to use them in production. The overall goal is to get you to understand how to use gRPC but also how gRPC works. First, you will get an introduction to the networking and Protobuf concepts you need to know before using gRPC. Then, you will see how to write unary and different stream APIs. And finally, in the rest of the book, you will learn about gRPC's features and how to create production-grade APIs.

Who this book is for

If you are a software engineer or architect and you ever struggled with writing APIs or you are looking for a more efficient alternative to existing APIs, this book is for you.

What this book covers

Chapter 1, *Networking Primer*, will teach you about the networking concepts behind gRPC.

Chapter 2, *Protobuf Primer*, will help you understand how Protobuf is essential for efficient communication.

Chapter 3, *Introduction to gPRC*, will give you a feel for why gRPC is more efficient than traditional REST APIs.

Chapter 4, *Setting Up a Project*, will mark the start of your journey into the gRPC world.

Chapter 5, *Types of gRPC Endpoints*, will describe how to write unary, server streaming, client streaming, and bidirectional streaming APIs.

Chapter 6, *Designing Effective APIs*, will outline the different trade-offs when designing gRPC APIs.

Chapter 7, *Out-of-the-Box Features*, will go through the major features that gRPC provides out of the box.

Chapter 8, *More Essential Features*, will explain how community projects can make your APIs even more powerful and secure.

Chapter 9, *Production-Grade APIs*, will teach you how to test, debug, and deploy your APIs.

To get the most out of this book

You must already be familiar with Go. Even though most of the time, you will only need basic coding skills, understanding Go's concurrency concepts will be handy.

Software/hardware covered in the book	Operating system requirements
Go 1.20.4	Windows, macOS, or Linux
Protobuf 23.2	Windows, macOS, or Linux
gRPC 1.55.0	Windows, macOS, or Linux
Buf 1.15.1	Windows, macOS, or Linux
Bazel 6.2.1	Windows, macOS, or Linux

If you are using the digital version of this book, we advise you to type the code yourself or access the code from the book's GitHub repository (a link is available in the next section). Doing so will help you avoid any potential errors related to the copying and pasting of code.

Download the example code files

You can download the example code files for this book from GitHub at `https://github.com/PacktPublishing/gRPC-Go-for-Professionals`. If there's an update to the code, it will be updated in the GitHub repository.

We also have other code bundles from our rich catalog of books and videos available at `https://github.com/PacktPublishing/`. Check them out!

Download the color images

We also provide a PDF file that has color images of the screenshots and diagrams used in this book. You can download it here: `https://packt.link/LEms7`.

Conventions used

There are a number of text conventions used throughout this book.

`Code in text`: Indicates code words in text, database table names, folder names, filenames, file extensions, pathnames, dummy URLs, user input, and Twitter handles. Here is an example: "With the plugin, we can use the `--validate_out` option in protoc."

A block of code is set as follows:

```
message AddTaskRequest {
  string description = 1;
```

```
  google.protobuf.Timestamp due_date = 2;
}
```

When we wish to draw your attention to a particular part of a code block, the relevant lines or items are set in bold:

```
proto/todo/v2
├── todo.pb.go
├── todo.pb.validate.go
├── todo.proto
└── todo_grpc.pb.go
```

Any command-line input or output is written as follows:

```
$ bazel run //:gazelle-update-repos
```

> **Tips or important notes**
> Appear like this.

Get in touch

Feedback from our readers is always welcome.

General feedback: If you have questions about any aspect of this book, email us at customercare@packtpub.com and mention the book title in the subject of your message.

Errata: Although we have taken every care to ensure the accuracy of our content, mistakes do happen. If you have found a mistake in this book, we would be grateful if you would report this to us. Please visit www.packtpub.com/support/errata and fill in the form.

Piracy: If you come across any illegal copies of our works in any form on the internet, we would be grateful if you would provide us with the location address or website name. Please contact us at copyright@packt.com with a link to the material.

If you are interested in becoming an author: If there is a topic that you have expertise in and you are interested in either writing or contributing to a book, please visit authors.packtpub.com.

Share Your Thoughts

Once you've read *gRPC Go for Professionals*, we'd love to hear your thoughts! Scan the QR code below to go straight to the Amazon review page for this book and share your feedback.

https://packt.link/r/1837638845

Your review is important to us and the tech community and will help us make sure we're delivering excellent quality content.

Download a free PDF copy of this book

Thanks for purchasing this book!

Do you like to read on the go but are unable to carry your print books everywhere? Is your eBook purchase not compatible with the device of your choice?

Don't worry, now with every Packt book you get a DRM-free PDF version of that book at no cost.

Read anywhere, any place, on any device. Search, copy, and paste code from your favorite technical books directly into your application.

The perks don't stop there, you can get exclusive access to discounts, newsletters, and great free content in your inbox daily

Follow these simple steps to get the benefits:

1. Scan the QR code or visit the link below

https://packt.link/free-ebook/9781837638840

2. Submit your proof of purchase
3. That's it! We'll send your free PDF and other benefits to your email directly

1

Networking Primer

Communication over networks is at the core of all of our modern technology and gRPC is one of the high-level frameworks that we can use to achieve efficient reception and transmission of data. As it is high level, it gives you abstractions for sending and receiving data without thinking about all the things that could go wrong when communicating over the wire. In this chapter, the goal is to understand, at a lower level (not the lowest), what happens when we send/receive messages in gRPC Go. This will help you get a sense of what's going on and, later on, when we talk about debugging and observability, you'll be able to grasp the concepts presented more easily.

In this chapter, we're going to cover the following main topics:

- HTTP/2
- RPC operations
- RPC types
- The life cycle of an RPC

Prerequisites

In this chapter, I will be using **Wireshark** (`https://www.wireshark.org/`) to analyze a gRPC call and get the necessary captures that illustrate the different concepts presented. As this is still early in this book, I do not expect you to reproduce all of this. However, if you plan to do a little bit more digging, you can install Wireshark and open the capture files provided in the Git repository (`https://github.com/PacktPublishing/gRPC-Go-for-Professionals`) under the `chapter1` directory.

To display these capture files, you can simply import them into Wireshark and apply a display filter to them. As we are interested specifically in HTTP/2 and gRPC payloads, and I was using port `50051` for communication, you can use the following filter: `tcp.port == 50051 and (grpc or http2)`.

Understanding HTTP/2

If you are reading this book, I'm going to assume that you have familiarity with HTTP/1.1 or that at least you have a sense of how to make traditional HTTP API calls over the network. I guess so because most of the APIs that we interact with, as developers, have concepts that were brought about by this protocol. I'm talking about concepts such as headers, which can provide metadata for a call; the body, which contains the main data; and actions such as GET, POST, UPDATE, and so on, which define what you intend to do with the data in the body.

HTTP/2 still has all of these concepts but improves efficiency, security, and usability in a few ways. The first advantage of HTTP/2 over plain old HTTP/1.1 is the compression down to binary. Before HTTP/2, everything sent over the network was pure text and it was up to the user to compress it or not. With version 2, every part of the HTTP semantic is translated down to binary, thus making it faster for computers to serialize and deserialize data between calls and thereby reducing the request/response payload size.

The second advantage that HTTP/2 has is a feature called server push. This is a feature that gives the server the ability to send multiple responses for only one call from the client. The overall goal here is to reduce the chatter between the server and client, and thus the total payload to reach the same end result. Without this feature, when a client wants to request a web page and all its resources, it has to do a request per resource. However, with the server push feature, the client can just send a request for a web page, and the server will return that web page, then return the CSS and potentially some JS script. This results in only one call from the client, instead of three.

Figure 1.1 – HTTP/2 server push

Another important efficiency aspect is the creation of a long-lived TCP connection instead of individual connections per request. In HTTP/0.9, every call is preceded by the creation of a TCP connection and succeeded by the closing of that connection. This is highly inefficient for today's use of the internet.

Then, HTTP/1.1 introduced the concept of KeepAlive, which permitted the reuse of a single TCP connection. However, this didn't mean that we could send interleaved packets to fulfill multiple requests concurrently; it meant that after finishing request one, we could reuse the same connection for request two.

This was probably fine in 1997 when the protocol was released, but nowadays we make more and more requests, and also bigger and bigger ones, and waiting for requests to finish before starting another one is not feasible. HTTP/2 solves this by creating a single long-lived connection that can handle multiple requests and responses as interleaved packets.

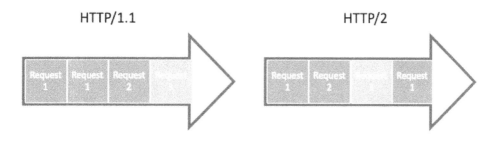

Figure 1.2 – HTTP/2 interleaved packets over the wire

What's presented here is obviously an oversimplification of the HTTP/2 protocol. It would take a book in itself to explain all the implementation details of the protocol. As we talk about gRPC, we mostly need to understand that in HTTP/2, we can send structured binary messages over the wire instead of text, we can have streams where the server can send multiple responses for one response, and finally, we do that in an efficient way because we only create one TCP connection and it will handle multiple requests and responses. However, it is also important to understand that gRPC has its own communication protocol on top of HTTP/2. This means that all the HTTP protocol improvements presented here are facilitators for communication. gRPC uses all of these in conjunction with four RPC operations.

RPC operations

Each interaction done with gRPC between the server and the client can be described as four RPC operations. These operations are composed in a way that creates complex high-level operations in the framework. Let us see these operations and then I will explain how a simple gRPC call uses them.

> **Important note**
>
> In this section, I'm going to use Wireshark's result for an RPC call. I will explain how to replicate what I did in this section later in the book. For now, I will just highlight what is important to notice in the dumps.

Send Header

The Send Header operation lets the server know that the client will send a request or lets the client know that the server will send a response. This acts as a switch between the server and client to let both sides know who needs to read and who needs to write.

By using Wireshark to analyze a simple gRPC call, we can observe the following header (simplified) being sent by the client in order to let the server know that it will send a request:

```
HyperText Transfer Protocol 2
    Stream: HEADERS, Stream ID: 1, Length 67, POST
        /greet.GreetService/Greet
        Flags: 0x04, End Headers
            00.0 ..0. = Unused: 0x00
            ..0. .... = Priority: False
            .... 0... = Padded: False
            .... .1.. = End Headers: True
            .... ...0 = End Stream: False
        Header: :method: POST
        Header: content-type: application/grpc
```

What is important to note in this header is that it mentions that the client wants to call HTTP POST on the /greet.GreetService/Greet route and then in the flags, it mentions that this is the end of the header data.

Then, later in the call, we will see the following header (simplified) sent by the server to let the client know that it will send the response:

```
HyperText Transfer Protocol 2
    Stream: HEADERS, Stream ID: 1, Length 14, 200 OK
        Flags: 0x04, End Headers
            00.0 ..0. = Unused: 0x00
            ..0. .... = Priority: False
            .... 0... = Padded: False
            .... .1.. = End Headers: True
            .... ...0 = End Stream: False
        Header: :status: 200 OK
        Header: content-type: application/grpc
```

And here, once again, we can see that this is a header and that this is the last one that will be sent. The main difference though, is that the server is telling the client that the request was handled properly, and it says that by sending a code 200.

Send Message

The Send Message operation is the operation that sends the actual data. This is the operation that matters the most to us as API developers. After sending the header, the client can send a message as a request and the server can send a message as a response.

By using Wireshark to analyze the same gRPC call as we did for Send Header, we can observe the following data (simplified) being sent by the client as a request:

```
GRPC Message: /greet.GreetService/Greet, Request
    0... .... = Frame Type: Data (0)
    .... ...0 = Compressed Flag: Not Compressed (0)
    Message Length: 9
    Message Data: 9 bytes
Protocol Buffers: /greet.GreetService/Greet, request
    Message: <UNKNOWN> Message Type
        Field(1):
            [Field Name: <UNKNOWN>]
            .000 1... = Field Number: 1
            .... .010 = Wire Type: Length-delimited (2)
            Value Length: 7
            Value: 436c656d656e74
```

What is important to note in this header is that it mentions that the client sends data on the /greet. GreetService/Greet route, which is the same as the one that was sent in the header. And then, we can see that we are sending Protocol Buffers data (more on that later) and that the binary value of that message is 436c656d656e74.

After, later in the call, just after the server header, we see the following data (simplified) sent by the server as a response:

```
GRPC Message: /greet.GreetService/Greet, Response
    0... .... = Frame Type: Data (0)
    .... ...0 = Compressed Flag: Not Compressed (0)
    Message Length: 15
    Message Data: 15 bytes
Protocol Buffers: /greet.GreetService/Greet, response
    Message: <UNKNOWN> Message Type
        Field(1):
            [Field Name: <UNKNOWN>]
            .000 1... = Field Number: 1
            .... .010 = Wire Type: Length-delimited (2)
            Value Length: 13
            Value: 48656c6c6f20436c656d656e74
```

And here, we can see that this is a message sent as a response to a call made on route `/greet`. `GreetService/Greet` and the binary value of that message is `48656c6c6f20436c656d656e74`.

Send Half Close

The `Send Half Close` operation closes either the input or the output of an actor. For example, in a traditional request/response setting, when the client is done sending the request, sending a `Half Close` closes the client stream. This is a little bit like `Send Header` in the sense that it is acting as a switch to let the server know that it is time to work.

Once again, if we look at the Wireshark transcript for the same gRPC call, we should be able to see that a header was set during the `Send Message` operations. We can observe the following data (simplified):

```
HyperText Transfer Protocol 2
    Stream: DATA, Stream ID: 1, Length 14
        Length: 14
        Type: DATA (0)
        Flags: 0x01, End Stream
            0000 .00. = Unused: 0x00
            .... 0... = Padded: False
            .... ...1 = End Stream: True
```

This time, we have a flag that denotes the end of the request. Notice however that here, we are sending a payload of type DATA. This is different than what we saw up until now because DATA is much lighter than a header. This is used for the Half Close because we simply want to send a boolean saying that the client is done.

Send Trailer

And finally, we have an operation for terminating an entire RPC. This is the `Send Trailer` operation. This operation also gives us more information about the call, such as the status code, error message, and so on. At this point in the book, we only need to know that this information is mostly necessary for handling API errors.

If we take a look at the same Wireshark call, we will have the following data (simplified):

```
HyperText Transfer Protocol 2
    Stream: HEADERS, Stream ID: 1, Length 24
        Length: 24
        Type: HEADERS (1)
        Flags: 0x05, End Headers, End Stream
            00.0 ..0. = Unused: 0x00
```

```
        ..0. .... = Priority: False
        .... 0... = Padded: False
        .... .1.. = End Headers: True
        .... ...1 = End Stream: True
   Header: grpc-status: 0
   Header: grpc-message:
```

Note that the trailer is basically a header. With this header, we will get more information about the call (`grpc-status` and `grpc-message`). And then we receive two flags – one saying that this is the end of the stream (in our case, request/response). And another one to say that this trailer ends here.

RPC types

Now that we know that there are four RPC operations, we can see how they are combined to create the different RPC types that gRPC provides. We will talk about the Unary, Server Streaming, Client Streaming, and Bidirectional RPC types. We will see that each type presented is a combination of the RPC operations presented earlier.

Unary

A unary RPC is an RPC that performs one request and returns one response. We already touched upon this in the previous section, but let us go ahead and make the process clearer.

As always, the first step is the client sending the initial header. This header will contain the information related to the RPC endpoint that we want to invoke. As of this point in the book, we simply need to know that this mostly includes the RPC route and the stream ID. The former is to let the server know which user code function it should call to process the request. The latter is a way to identify on which stream the data should be sent. This is because we can have multiple streams going on at the same time.

Since the server is now aware of the fact that the client will send a request, the client can now send a message. This message will contain the actual request payload. In our case, we are going to only send Protocol Buffers encoded data, but be aware that you can send any kind of data with gRPC.

After that, because we are in a unary setting, the client is done sending the request. As we know, the client itself should now send a Half Close. This is saying to the server *I'm done, please send me the response*.

At this point, the server will do similar work. As shown in the next figure, it will send a header, send the response as a Protobuf encoded message, and end the RPC. However, as we know, the server will not send a Half Close; it will send the Trailer. This is some data that says whether or not the call was successful, has an optional error message, and some other key-value pairs that we can add through the user code.

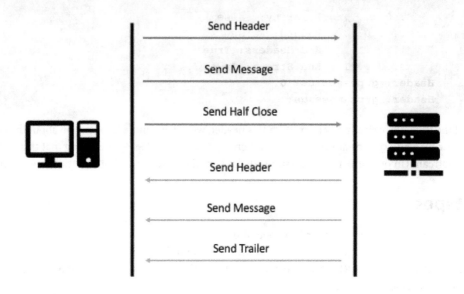

Figure 1.3 – Unary RPC flow

Server streaming

A server streaming RPC is an RPC that performs one request and returns one or more responses. This RPC type is useful for situations where the client is expecting to get updates from the server. For example, we could have a client display the stock prices for selected companies. With server streaming, the client could subscribe and the server could send different prices over time.

In this situation, nothing changes for the client. It will send the header, the message, and the Half Close. However, on the server side, we are going to interleave HTTP data messages and data payloads.

As shown in the following figure, the server will first send its header. This is customary when an actor wants to let the other one know that it will send messages. After that, as mentioned above, the server will alternate between sending HTTP data messages and Protobuf payloads. The first data messages will look like this (simplified):

```
HyperText Transfer Protocol 2
    Stream: DATA, Stream ID: 1, Length 30
        Length: 30
        Type: DATA (0)
        Flags: 0x00
            0000 .00. = Unused: 0x00
            .... 0... = Padded: False
            .... ...0 = End Stream: False
```

This says that a message will be sent. This is a lightweight header. And later, once we arrive at the last message to be sent, the server will finish the RPC with the trailer, and at that point, the client will know that the server is done sending responses.

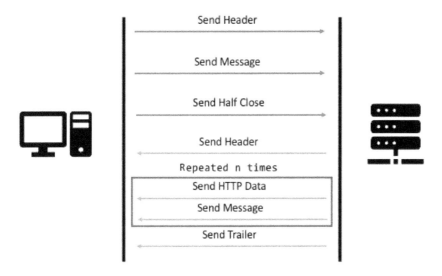

Figure 1.4 – Server streaming flow

Client streaming

A client streaming RPC is similar to server streaming, but this time, the client can send one or more requests and the server returns one response. This is useful in situations where the client sends real-time information to the server. For example, this might be useful for microcontrollers to send data coming from some kind of sensor and update the server on the current state of what the sensor is measuring.

Client streaming is similar to server streaming. As you can see in the following figure, the client will do what the server did in server streaming. This means that the client will interleave HTTP data messages, which are like the ones previously mentioned about server streaming, with the Protobuf messages. And in the end, when the client is done, it will simply send a Half Close.

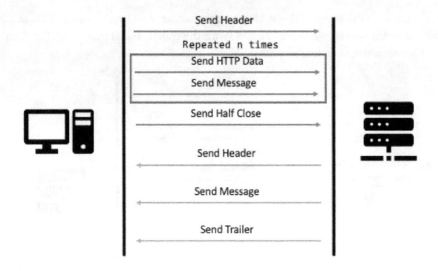

Figure 1.5 – Client streaming flow

Bidirectional streaming

By now, you might have guessed that bidirectional streaming is a mix of server streaming and client streaming. A client can send one or more requests and the server returns one or more responses. This is especially useful when one of the actors needs feedback on its data. For example, if you have an app to find taxis, it might not be enough for the server to send updates about cabs. The user might also walk toward their destination in the hope of catching a taxi on the road. Thus, the server also needs to know the user's location.

Bidirectional streaming is less predictable than client and server streaming. This is because there is no defined order in which each actor will send messages. The server could give a response per request or any number of requests. Thus, for this section, let us pretend that we are working with a server that returns a response per request.

In this case, as you can see in the following figure, the client will send a header and a message. Then, the server will send its header and message. And after that, we will get data and a message per actor. And finally, we will get the Half Close from the client and the Trailer from the server.

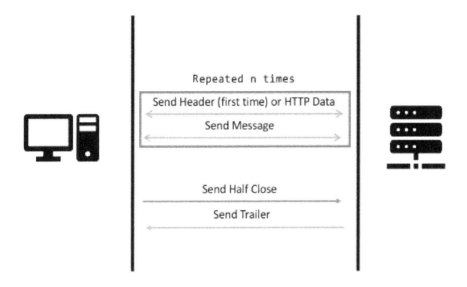

Figure 1.6 – Bidirectional streaming flow

The life cycle of an RPC

Now that we understand the basic RPC operations that can be executed in gRPC and the different types of RPC, we can take a look at the life cycle of an RPC. In this section, we are going to go top-down by first explaining the overall idea of what is happening when a client sends a request and the server receives it, sends a response, and the client receives it. And after that, we will go a bit deeper and talk about three stages:

1. The connection – What happens when a client connects to a server?

2. The client side – What happens when a client sends a message?

3. The server side – What happens when a server receives a message?

> **Important note**
>
> gRPC has multiple implementations in different languages. The original one was in C++ and some implementations are just wrappers around the C++ code. However, gRPC Go is a standalone implementation. This means that it was implemented from scratch in Go and doesn't wrap up the C++ code. As such, in this section, we are going to talk specifically about gRPC Go, and this might prove to be implemented differently in other implementations.

Before going into too much detail, let's start with a bird's-eye view by defining some concepts. The first thing that we need to be clear on is that gRPC is driven by generated code in the user code. This means that we interact with only a few points of the gRPC API and we mostly deal with code that was generated based on our Protocol Buffer service definition. Do not worry too much about that yet; we are going to cover that in the last section.

The second important concept is the concept of transport. The transport can be seen as the manager of the connection between the actors and it sends/receives raw bytes over the network. It contains a read-write stream that is designed to be able to read and write over the network in any order. And the most important aspect in our case is that we can call the read on an io.Reader and we can call write on an io.Writer.

Finally, the last thing to clarify is that a client and a server are very similar. All the functions called on the client will be called on the server too. They will simply be called on different objects (for instance, ClientTransport and ServerTransport).

Now, that we understand all of this, we can take a look at a visual representation of the life cycle of an RPC.

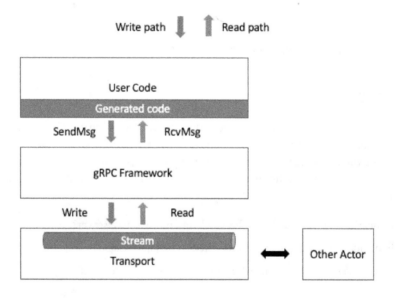

Figure 1.7 – Bird's-eye view of the RPC life cycle

We can see that we can simply define only one generic actor and that it will represent both a server and a client. Then, we can see that the generated code will interact directly with the gRPC framework by calling a function called SendMsg.

This is designed, as its name suggests, to send data over the network. This SendMsg function will call a lower-level function called Write. This is the function provided by the io.Writer in the

Transport. Once this is done, the other actor will read on an io.Reader, then the RcvMsg function, and finally, the user code will receive the data.

We are now going to go a bit deeper into the important parts of gRPC communication. And as for every kind of transmission over the wire, the client needs to connect to a server, we are going to start with the specifics of a connection.

The connection

To create a connection, the client code will call a function called Dial with a target URI and some options as parameters. When the Dial request is received, the gRPC framework will parse the target address according to RFC 3986 and it will create a Resolver depending on the scheme of the URI. So, for example, if we use the dns:// scheme, which is the default scheme that gRPC uses when we omit the scheme in the URI or if the scheme provided is unknown, gRPC will create a dnsResolver object.

dnsResolver Then the resolver will do its job, which is resolving the hostname and returning a list of addresses that can be connected to. With these addresses, gRPC will create a load balancer based on the configuration the user passed in the Dial options. The framework provides two load balancers by default:

- Pick first (the default), which connects to the first address it can connect to, and sends all the RPCs to it

- Round robin, which connects to all the addresses and sends an RPC to each backend one at a time and in order

As we can see, the goal of the load balancer is to find out on which address(es) the client should create a connection(s). So, it will return a list of addresses to which gRPC should connect, and then gRPC will create a channel, which is an abstraction for the connection used by RPCs, and subchannels, which are abstractions for connections that the load balancer can use to direct the data to one or more backends.

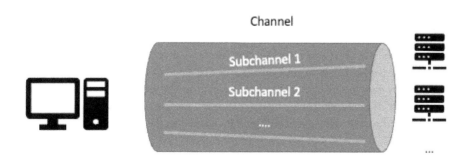

Figure 1.8 – Channels versus subchannels

In the end, the user code will receive a ClientConn object that will be used for closing the connection, but most importantly, to create a client object that is defined in the generated code and on which we will be able to call the RPC endpoints. The last thing to note here is that, by default, the overall process is non-blocking. This means that gRPC does not wait for the connections to be established to return the ClientConn object.

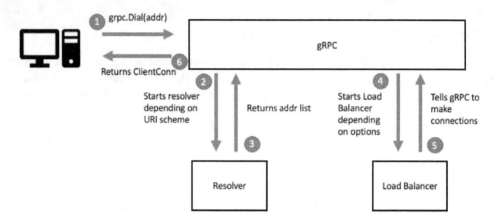

Figure 1.9 – Summary of an RPC connection

The client side

Now that we have a connection, we can start thinking about making requests. For now, let us say that we have generated code and that it has a Greet RPC endpoint. And for our current purpose, it is not important what it is doing; it is just an API endpoint.

To send a request, the user code will simply call the Greet endpoint. This will trigger a function called NewStream in the gRPC framework. The name of that function is a little bit of a misnomer because here a stream does not necessarily represent a streaming RPC. In fact, whether you are doing a streaming RPC or not, it will be called and it will create a ClientStream object. So here, Stream is roughly equivalent to an abstraction for all RPCs.

During the creation of that ClientStream, the gRPC framework will perform two actions. The first one is that it will call the load balancer to get a subchannel that can be used. This is done depending on the load balancer policy chosen during the connection creation. The second action is to interact with the transport. The gRPC framework will create the ClientTransport, which contains the read-write stream to send and receive data, and it will send the header to the server to initiate an RPC call.

Once this is done, the gRPC framework will simply return `ClientStream` to the generated code and the generated code will simply encapsulate it with another object to provide the user code with a smaller set of functions to be called (for example, `Send`, `Recv`, and so on).

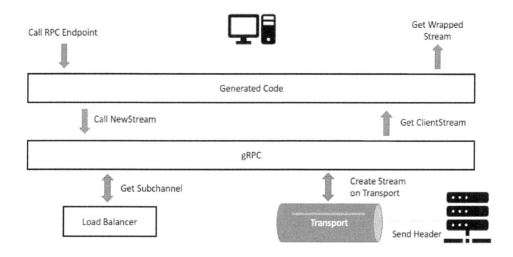

Figure 1.10 – Summary of client-side communication

The server side

Naturally, after sending a request, we expect a response from the server. As we know by now, the client sent a header to initiate an RPC call. This header will be handled by `ServerTransport`. The server is now aware that a client wants to send a request for the `Greet` RPC endpoint.

With that, the transport will send a `transport.Stream` object to the gRPC framework. Then, again, this stream will be thinly wrapped in a `ServerStream` object and passed to the generated code. At this point, the generated code is aware of which user code function to call. It is aware of that because the user code registers functions to specific RPC endpoints.

And that is it, the server will do the computation of the data received and it will simply return a response on the corresponding transport to the client. The `ClientTransport` will read that and return the response to the user code.

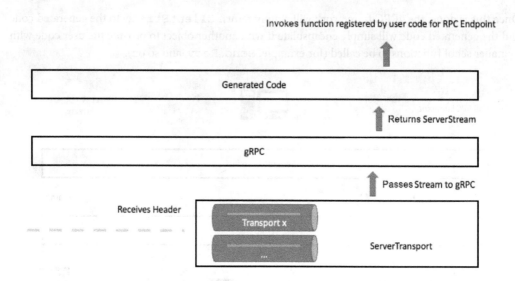

Figure 1.11 – Summary of server-side communication

Summary

All that knowledge might be overwhelming right now, but do not worry, you do not need to remember all the names of the objects presented to understand how gRPC works. The point of this chapter is more about giving you a sense of the different actors involved in the process of making a connection and sending/receiving data.

We saw that we have four RPC operations that can be performed by the client and/or the server. Each actor sends a header to indicate it is its turn to send data, then they send messages, and finally, each of them has a special operation to indicate that it is done with sending messages.

After that, we saw how gRPC creates a connection between the server and the client. This is done with the help of the resolver, which finds IP addresses depending on the address we try to connect to, and with the load balancer, which helps gRPC work out which subchannels to send the data to.

Then, we talked about channels and subchannels. We saw how they are created by the client to connect to the server. And finally, we saw that the server will receive data and call some code that the user code registered for an RPC endpoint.

In the next chapter, we will introduce Protocol Buffers and how they relate to gRPC.

Quiz

1. What is the RPC operation that tells the server that the client is ready to send a request?

    A. `Send Trailer`

    B. `Send Message`

    C. `Send Header`

2. What is the RPC operation that tells the client that the server is done returning response(s)?

    A. `Send Half-Close`

    B. `Send Trailer`

    C. `Send Header`

3. Which RPC type can be used for downloading information by chunks in one request from the client side?

    A. Server streaming

    B. Client streaming

    C. Bidirectional streaming

    D. Unary

4. Which RPC type is the equivalent of a traditional HTTP/1.1 request?

    A. Server streaming

    B. Client streaming

    C. Bidirectional streaming

    D. Unary

5. What is a channel?

    A. An abstraction used by RPCs for representing a connection to any available server discovered by the load balancer.

    B. An abstraction used by the load balancer for representing a connection to a specific server.

    C. Both of the above

6. What is a subchannel?

    A. An abstraction used by RPCs for representing a connection to any available server discovered by the load balancer.

    B. An abstraction used by the load balancer for representing a connection to a specific server.

    C. Both of the above

7. When receiving the `ClientConn` object from `grpc.Dial`, can you be sure that the client has established a connection with the server?

    A. Yes

    B. No

Answers

1. C

2. B

3. A

4. D

5. A

6. B

7. B

2

Protobuf Primer

As we now understand the basic networking concepts behind gRPC, we can touch upon another pillar in the construction of your gRPC APIs. This pillar is **Protocol Buffers**, more commonly known as **Protobuf**. It is an important part of the communication process because, as we saw in the previous chapter, every message is encoded into binary, and this is exactly what Protobuf is doing for us in gRPC. In this chapter, the goal is to understand what Protobuf is and why it is needed for high-efficiency communication. Finally, we are going to look at some details concerning the serialization and deserialization of messages.

In this chapter, we're going to cover the following main topics:

- Protobuf is an **Interface Description Language** (IDL)
- Serialization/deserialization
- Protobuf versus JSON
- Encoding details
- Common types
- Services

Prerequisites

You can find the code for this chapter at `https://github.com/PacktPublishing/gRPC-Go-for-Professionals/tree/main/chapter2`. In this chapter, we are going to discuss how Protocol Buffers serializes and deserializes data. While this can be done by writing code, we are going to stay away from that in order to learn how to use the protoc compiler to debug and optimize our Protobuf schemas. Thus, if you want to reproduce the examples specified, you will need to download the protoc compiler from the Protobuf GitHub *Releases* page (`https://github.com/protocolbuffers/protobuf/releases`). The easiest way to get started is to download the binary releases. These releases are named with this convention: `protoc-${VERSION}-${OS}-{ARCHITECTURE}`. Uncompress the zip file and follow the `readme.txt` instructions (note: we do intend to use Well-Known Types in the future so make sure you also install the includes). After that, you should be able to run the following command:

```
$ protoc --version
```

Finally, as always, you will be able to find the companion code in the GitHub repository under the folder for the current chapter (`chapter2`).

Protobuf is an IDL

Protobuf is a language. More precisely, it is an IDL. It is important to make such a distinction because, as we will see more in detail later, in Protobuf, we do not write any logic the way we do in a programming language, but instead, we write data schemas, which are contracts to be used for serialization and are to be fulfilled by deserialization. So, before explaining all the rules that we need to follow when writing a `.proto` file and going through all the details about serialization and deserialization, we need to first get a sense of what an IDL is and what is the goal of such a language.

An IDL, as we saw earlier, is an acronym for *Interface Description Language*, and as we can see, the name contains three parts. The first part, **Interface**, describes a piece of code that sits in between two or more applications and hides the complexity of implementation. As such, we do not make any assumptions about the hardware on which an application is running, the OS on which it runs, and in which programming language it is written. This interface is, by design, hardware-, OS-, and language-agnostic. This is important for Protobuf and several other serialization data schemas because it lets developers write the code once and it can be used across different projects.

The second part is **Description**, and this sits on top of the concept of Interface. Our interface is describing what the two applications can expect to receive and what they are expected to send to each other. This includes describing some types and their properties, the relationship between these types, and the way these types are serialized and deserialized. As this may be a bit abstract, let us look at an example in Protobuf. If we wanted to create a type called `Account` that contains an ID, a username, and the rights this account has, we could write the following:

```
syntax = "proto3";

enum AccountRight {
  ACCOUNT_RIGHT_UNSPECIFIED = 0;
  ACCOUNT_RIGHT_READ = 1;
  ACCOUNT_RIGHT_READ_WRITE = 2;
  ACCOUNT_RIGHT_ADMIN = 3;
}
message Account {
  uint64 id = 1;
```

```
  string username = 2;
  AccountRight right = 3;
}
```

If we skip some of the details that are not important at this stage, we can see that we define the following:

- An enumeration listing all the possible rights and an extra role called ACCOUNT_RIGHT_UNSPECIED

- A message (equivalent to a class or struct) listing the three properties that an Account type should have

Again, without looking at the details, it is readable, and the relationship between Account and AccountRight is easy to understand.

Finally, the last part is **Language**. This is here to say that, as with every language—computer ones or not—we have rules that we need to follow so that another human, or a compiler, can understand our intent. In Protobuf, we write our code to please the compiler (protoc), and then it does all the heavy lifting for us. It will read our code and generate code in the language that we need for our application, and then our user code will interact with the generated code. Let us look at a simplified output of what the Account type defined previously would give in Go:

```
type AccountRight int32
const (
  AccountRight_ACCOUNT_RIGHT_UNSPECIFIED AccountRight = 0
  AccountRight_ACCOUNT_RIGHT_READ AccountRight = 1
  AccountRight_ACCOUNT_RIGHT_READ_WRITE AccountRight = 2
  AccountRight_ACCOUNT_RIGHT_ADMIN AccountRight = 3
)

type Account struct {
  Id uint64 `protobuf:"varint,1,…`
  Username string `protobuf:"bytes,2,…`
  Right AccountRight `protobuf:"varint,3,…`
}
```

In this code, there are important things to notice. Let us break this code into pieces:

```
type AccountRight int32
const (
  AccountRight_ACCOUNT_RIGHT_UNSPECIFIED AccountRight = 0
  AccountRight_ACCOUNT_RIGHT_READ AccountRight = 1
  AccountRight_ACCOUNT_RIGHT_READ_WRITE AccountRight = 2
  AccountRight_ACCOUNT_RIGHT_ADMIN AccountRight = 3
)
```

Our `AccountRight` enum is defined as constants with values of type `int32`. Each enum variant's name is prefixed with the name of the enum, and each constant has the value that we set after the equals sign in the Protobuf code. These values are called field tags, and we will introduce them later in this chapter.

Now, take a look at the following code:

```
type Account struct {
  Id uint64 `protobuf:"varint,1,…`
  Username string `protobuf:"bytes,2,…`
  Right AccountRight `protobuf:"varint,3,…`
}
```

Here, we have our `Account` message transpiled to a struct with `Id`, `Username`, and `Right` exported fields. Each of these fields has a type that is converted from a Protobuf type to a Golang type. In our example here, Go types and Protobuf types have the exact same names, but it is important to know that in some cases, the types will translate differently. Such an example is `double` in Protobuf, which will translate to `float64` for Go. Finally, we have the field tags, referenced in the metadata following the field. Once again, their meaning will be explained later in this chapter.

So, to recapitulate, an IDL is a piece of code sitting between different applications and describing objects and their relationships by following certain defined rules. This IDL, in the case of Protobuf, will be read, and it will be used to generate code in another language. And after that, this generated code will be used by the user code to serialize and deserialize data.

Serialization and deserialization

Serialization and deserialization are two concepts that are used in many ways and in many kinds of applications. This section is going to discuss these two concepts in the context of Protobuf. So, even if you feel confident about your understanding of these two notions, it is important to get your head straight and understand them properly. Once you do, it will be easier to deal with the *Encoding details* section where we are going to delve deeper into how Protobuf serializes and deserializes data under the hood.

Let us start with serialization and then let us touch upon deserialization, which is just the opposite process. The goal of serialization is to store data, generally in a more compact or readable representation, to use it later. For Protobuf, this serialization happens on the data that you set in your generated code's objects. For example, if we set the `Id`, `Username`, and `Right` fields in our `Account` struct, this data will be what Protobuf will work on. It will turn each field into a binary representation with different algorithms depending on the field type. And after that, we use this in-memory binary to either send data over the network (with gRPC, for example) or store it in more persistent storage.

Once it is time for us to use this serialized data again, Protobuf will perform deserialization. This is the process of reading the binary created earlier and populating the data back into an object in your favorite programming language to be able to act on it. Once again, Protobuf will use different algorithms depending on the type of data to read the underlying binary and know how to set or not set each of the fields of the object in question.

To summarize, Protobuf performs binary serialization to make data more compact than other formats such as XML or JSON. To do so, it will read data from the different fields of the generated code's object, turn it into binary with different algorithms, and then when we finally need the data, Protobuf will read the data and populate the fields of a given object.

Protobuf versus JSON

If you've already worked on the backend or even frontend, there is a 99.99 percent chance that you've worked with JSON. This is by far the most popular data schema out there and there are reasons why it is the case. In this section, we are going to discuss the pros and cons of both JSON and Protobuf and we are going to explain which one is more suitable for which situation. The goal here is to be objective because as engineers, we need to be to choose the right tool for the right job.

As we could write chapters about the pros and cons of each technology, we are going to reduce the scope of these advantages and disadvantages to three categories. These categories are the ones that developers care the most about when developing applications, as detailed here:

- **Size of serialized data**: We want to reduce the bandwidth when sending data over the network
- **Readability of the data schema and the serialized data**: We want to be able to have a descriptive schema so that newcomers or users can quickly understand it, and we want to be able to visualize the data serialized for debugging or editing purposes
- **Strictness of the schema**: This quickly becomes a requirement when APIs grow, and we need to ensure the correct type of data is being sent and received between different applications

Serialized data size

In serialization, the Holy Grail is, in a lot of use cases, reducing the size of your data. This is because most often, we want to send that data to another application across the network, and the lighter the payload, the faster it should arrive on the other side. *In this space, Protobuf is the clear winner against JSON.* This is the case because JSON serializes to text whereas Protobuf serializes to binary and thus has more room to improve how compact the serialized data is. An example of that is numbers. If you set a number to the `id` field in JSON, you would get something like this:

```
{ id: 123 }
```

First, we have some boilerplate with the braces, but most importantly we have a number that takes three characters, or three bytes. In Protobuf, if we set the same value to the same field, we would get the hexadecimal shown in the following callout.

> **Important note**
>
> In the `chapter2` folder of the companion GitHub repository, you will find the files need to reproduce all the results in this chapter. With protoc, we will be able to display the hexadecimal representation of our serialized data. To do that, you can run the following command:
>
> Linux/Mac: `cat ${INPUT_FILE_NAME}.txt | protoc --encode=${MESSAGE_NAME} ${PROTO_FILE_NAME}.proto | hexdump -C`
>
> Windows (PowerShell): `(Get-Content ${INPUT_FILE_NAME}.txt | protoc --encode=${MESSAGE_NAME} ${PROTO_FILE_NAME}.proto) -join "`n" | Format-Hex`
>
> For example:
>
> ```
> $ cat account.txt | protoc --encode=Account account.proto | hexdump -C
> 00000000  08 7b                                             |.{|
> 00000002
> ```

Right now, this might look like magic numbers, but we are going to see in the next section how it is encoded into two bytes. Now, two bytes instead of three might look negligible but imagine this kind of difference at scale, and you would have wasted millions of bytes.

Readability

The next important thing about data schema serialization is readability. However, readability is a little bit too broad, especially in the context of Protobuf. As we saw, as opposed to JSON, Protobuf separates the schema from the serialized data. We write the schema in a `.proto` file and then the serialization will give us some binary. In JSON, the schema is the actual serialized data. So, to be clearer and more precise about readability, let us split readability into two parts: the readability of the schema and the readability of the serialized data.

As for the readability of the schema, this is a matter of preference, but there are a few points that make Protobuf stand out. The first one of them is that Protobuf can contain comments, and this is nice to have for extra documentation describing requirements. JSON does not allow comments in the schema, so we must find a different way to provide documentation. Generally, it is done with GitHub wikis or other external documentation platforms. This is a problem because this kind of documentation quickly becomes outdated when the project and the team working on it get bigger. A simple oversight and your documentation do not describe the real state of your API. With Protobuf, it is still possible to have outdated documentation, but as the documentation is closer to the code, it provides more incentive and awareness to change the related comment.

The second feature that makes Protobuf more readable is the fact that it has explicit types. JSON has types but they are implicit. You know that a field contains a string if its value is surrounded by double quotes, a number when the value is only digits, and so on. In Protobuf, especially for numbers, we get more information out of types. If we have an `int32` type, we can obviously know that this is a number, but on top of that, we know that it can accept negative numbers and we are able to know the range of numbers that can be stored in this field. Explicit types are important not only for security (more on that later) but also for letting the developer know the details of each field and letting them describe accurately their schemas to fulfill the business requirements.

For readability of the schema, I think we can agree that Protobuf is the winner here because it can be written as self-documenting code and we get explicit types for every field in objects.

As for the readability of serialized data, JSON is the clear winner here. As mentioned, JSON is both the data schema and the serialized data. What you see is what you get. Protobuf, however, serializes the data to binary, and it is way harder to read that, even if you know how Protobuf serializes and deserializes data. In the end, this is a trade-off between readability and serialized data size here. Protobuf will outperform JSON on serialized data and is way more explicit on the readability of the data schema. However, if you need human-readable data that can be edited by hand, Protobuf is not the right fit for your use case.

Schema strictness

Finally, the last category is the strictness of the schema. This is usually a nice feature to have when your team and your project scale because it ensures that the schema is correctly populated, and for a certain target language, it shortens the feedback loop for the developers.

Schemas are always valid ones because every field has an explicit type that can only contain certain values. We simply cannot pass a string to a field that was expecting a number or a negative number to a field that was expecting a positive number. This is enforced in the generated code by either runtime checks for dynamic languages or at compile time for typed languages. In our case, since Go is a typed language, we will have compile-time checks.

And finally, in typed languages, a schema shortens the feedback loop because instead of having a runtime check that might or might not trigger an error, we simply have a compilation error. This makes our software more reliable, and developers can feel confident that if they were able to compile, the data set into the object would be valid.

In pure JSON, we cannot ensure that our schema is correct at compile time. Most often, developers will add extra configurations such as JSON Schema to have this kind of assurance at runtime. This adds complexity to our project and requires every developer to be disciplined because they could simply go about their code without developing the schema. In Protobuf, we do schema-driven development. The schema comes first, and then our application revolves around the generated types. Furthermore, we have assurance at compile time that the values that we set are correct and we do not need to replicate the setup to all our microservices or subprojects. In the end, we spend less time on configuration and we spend more time thinking about our data schemas and the data encoding.

Encoding details

Up until now, we talked a lot about "algorithms"; however, we did not get too much into the specifics. In this section, we are going to see the major algorithms that are behind the serialization and deserialization processes in Protobuf. We are first going to see all the types that we can use for our fields, then with that, we are going to divide them into three categories, and finally, we are going to explain which algorithm is used for each category.

In Protobuf, types that are considered simple and that are provided by Protobuf out of the box are called **scalar types**. We can use 15 of such types, as listed here:

- `int32`
- `int64`
- `uint32`
- `uint64`
- `sint32`
- `sint64`
- `fixed32`
- `fixed64`
- `sfixed32`
- `sfixed64`
- `double`
- `float`
- `string`
- `bytes`
- `bool`

And out of these 15 types, 10 are for integers (the 10 first ones). These types might be intimidating at first, but do not worry too much about how to choose between them right now; we are going to discuss that throughout this section. The most important thing to understand right now is that two-thirds of the types are for integers, and this shows what Protobuf is good at—encoding integers.

Now that we know the scalar types, let us separate these types into three categories. However, we are not here to make simple categories such as numbers, arrays, and so on. We want to make categories that are related to the Protobuf serialization algorithms. In total, we have three: fixed-size numbers, variable-size integers (varints), and length-delimited types. Here is a table with each category populated:

Fixed-size numbers	Varints	Length-delimited types
fixed32	int32	string
fixed64	int64	bytes
sfixed32	uint32	
sfixed64	uint64	
double	bool	
float		

Let's go through each now.

Fixed-size numbers

The easiest one to understand for developers who are used to typed languages is fixed-size numbers. If you worked with lower-level languages in which you tried to optimize storage space, you know that we can, on most hardware, store an integer in 32 bits (4 bytes) or in 64 bits (8 bytes). fixed32 and fixed64 are just binary representations of a normal number that you would have in languages that give you control over the storage size of your integers (for example, Go, C++, Rust, and so on). If we serialize the number 42 into a fixed32 type, we will have the following:

```
$ cat fixed.txt | protoc --encode=Fixed32Value
  wrappers.proto | hexdump -C
00000000  0d 2a 00 00 00                                    |.*...|
00000005
```

Here, 2a is 42, and 0d is a combination of the field tag and the type of the field (more about that later in this section). In the same manner, if we serialize 42 in a fixed64 type, we will have the following:

```
$ cat fixed.txt | protoc --encode=Fixed64Value
  wrappers.proto | hexdump -C
00000000  09 2a 00 00 00 00 00 00  00                       |.*.......|
00000009
```

And the only thing that changed is the combination between the type of the field and the field tag (09). This is mostly because we changed the type to 64-bit numbers.

Two other scalar types that are easy to understand are float and double. Once again, Protobuf produces the binary representation of these types. If we encode 42.42 as float, we will get the following output:

```
$ cat floating_point.txt | protoc --encode=FloatValue
  wrappers.proto | hexdump -C
00000000  0d 14 ae 29 42                                    |...)B|
00000005
```

In this case, this is a little bit more complicated to decode, but this is simply because float numbers are encoded differently. If you are interested in this kind of data storage, you can look at the *IEEE Standard for Floating-Point Arithmetic* (*IEEE 754*), which explains how a float is formed in memory. What is important to note here is that floats are encoded in 4 bytes, and in front, we have our tag + type. And for a `double` type with a value of `42.42`, we will get the following:

```
$ cat floating_point.txt | protoc --encode=DoubleValue
  wrappers.proto | hexdump -C
00000000  09 f6 28 5c 8f c2 35 45  40           |..(\..5E@|
00000009
```

This is encoded in 8 bytes and the tag + type. Note that the tag + type also changed here because we are in the realm of 64-bit numbers.

Finally, we are left with `sfixed32` and `sfixed64`. We did not mention it earlier, but `fixed32` and `fixed64` are unsigned numbers. This means that we cannot store negative numbers in fields with these types. `sfixed32` and `sfixed64` solve that. So, if we encode –42 in a `sfixed32` type, we will have the following:

```
$ cat sfixed.txt | protoc --encode=SFixed32Value
  wrappers.proto | hexdump -C
00000000  0d d6 ff ff ff                         |.....|
00000005
```

This is obtained by taking the binary for 42, flipping all the bits (1's complement), and adding one (2's complement). Otherwise, if you serialize a positive number, you will have the same binary as the `fixed32` type. Then, if we encode –42 in a field with type `sfixed64`, we will get the following:

```
$ cat sfixed.txt | protoc --encode=SFixed64Value
  wrappers.proto | hexdump -C
00000000  09 d6 ff ff ff ff ff ff  ff           |.........|
00000009
```

This is like the `sfixed32` type, only the tag + type was changed.

To summarize, fixed integers are simple binary representations of integers that resemble how they are stored in most computers' memory. As their name suggests, their serialized data will always be serialized into the same number of bytes. For some use cases, this is fine to use such representations; however, in most cases, we would like to reduce the number of bits that are just here for padding. And in these use cases, we will use something called varints.

Varints

Now that we have seen fixed integers, let us move to another type of serialization for numbers: variable-length integers. As its name suggests, we will not get a fixed number of bytes when serializing an integer.

To be more precise, the smaller the integer, the smaller the number of bytes it will be serialized into, and the bigger the integer, the larger the number of bytes. Let us look at how the algorithm works.

In this example, let us serialize the number 300. To start, we are going to take the binary representation of that number:

```
100101100
```

With this binary, we can now split it into groups of 7 bits and pad with zeros if needed:

```
0000010
0101100
```

Now, since we lack 2 more bits to create 2 bytes, we are going to add 1 as the **most significant bit** (**MSB**) for all the groups except the first one, and we are going to add 0 as the MSB for the first group:

```
00000010
10101100
```

These MSBs are continuation bits. This means that, when we have 1, we still have 7 bits to read after, and if we have 0, this is the last group to be read. Finally, we put this number into little-endian order, and we have the following:

```
10101100 00000010
```

Or, we would have AC 02 in hexadecimal. Now that we have serialized 300 into AC 02, and keeping in mind that deserialization is the opposite of serialization, we can deserialize that data. We take our binary representation for AC 02, drop the continuation bits (MSBs), and we reverse the order of bytes. In the end, we have the following binary:

```
100101100
```

This is the same binary we started with. It equals 300.

Now, in the real world, you might have larger numbers. For a quick reference on positive numbers, here is a list of the thresholds at which the number of bytes will increase:

Threshold value	Byte size
0	0
1	1
128	2
16,384	3
2,097,152	4
268,435,456	5
34,359,738,368	6

4,398,046,511,104	7
562,949,953,421,312	8
72,057,594,037,927,936	9
9,223,372,036,854,775,807	9

An astute reader might have noticed that having a varint is often beneficial, but in some cases, we might encode our values into more bytes than needed. For example, if we encode 72,057,594,037,927,936 into an int64 type, it will be serialized into 9 bytes, while with a fixed64 type, it will be encoded into 8. Furthermore, a problem coming from the encoding that we just saw is that negative numbers will be encoded into a large positive number and thus will be encoded into 9 bytes. That begs the following question: *How can we efficiently choose between the different integer types?*

How to choose?

The answer is, as always, it depends. However, we can be systematic in our choices to avoid many errors. We mostly have three choices that we need to make depending on the data we want to serialize:

- The range of numbers needed
- The need for negative numbers
- The data distribution

The range

By now, you might have noticed that the 32 and 64 suffixes on our types are not always about the number of bits into which our data will be serialized. For varints, this is more about the range of numbers that can be serialized. These ranges are dependent on the algorithm used for serialization.

For fixed, signed, and variable-length integers, the range of numbers is the same as the one developers are used to with 32 and 64 bits. This means that we get the following:

```
[-2^(NUMBER_OF_BITS - 1), 2^(NUMBER_OF_BITS - 1) - 1]
```

Here, NUMBER_OF_BITS is either 32 or 64 depending on the type you want to use.

For unsigned numbers (uint)—this is again like what developers are expecting—we will get the following range:

```
[0, 2 * 2^(NUMBER_OF_BITS - 1) - 1]
```

The need for negative numbers

In the case where you simply do not need negative numbers (for example, for IDs), the ideal type to use is an unsigned integer (uint32, uint64). This will prevent you from encoding negative numbers, it will have twice the range in positive numbers compared to signed integers, and it will serialize using the varint algorithm.

And another type that you will potentially work with is the one for signed integers (`sint32`, `sint64`). We won't go into details about how to serialize them, but the algorithm transforms any negative number into a positive number (ZigZag encoding) and serializes the positive number with the varint algorithm. This is more efficient for serializing negative numbers because instead of being serialized as a large positive number (9 bytes), we take advantage of the varint encoding. However, this is less efficient for serializing positive numbers because now we interleave the previously negative numbers and the positive numbers. This means that for the same positive number, we might have different amounts of encoding bytes.

The data distribution

Finally, one thing that is worth mentioning is that encoding efficiency is highly dependent on your data distribution. You might have chosen some types depending on some assumptions, but your actual data might be different. Two common examples are choosing an `int32` or `int64` type because we expect to have few negative values and choosing an `int64` type because we expect to have few very big numbers. Both situations might result in significant inefficiencies because, in both cases, we might get a lot of values serialized into 9 bytes.

Unfortunately, there is no way of deciding the type that will always perfectly fit the data. In this kind of situation, there is nothing better than doing experiments on real data that is representative of your whole dataset. This will give you an idea of what you are doing correctly and what you are doing wrong.

Length-delimited types

Now that we've seen all the types for numbers, we are left with the length-delimited types. These are the types, such as string and bytes, from which we cannot know the length at compile time. Think about these as dynamic arrays.

To serialize such a dynamic structure, we simply prefix the raw data with the length of that data that is following. This means that if we have a string of length 10 and content "0123456789", we will have the following sequence of bytes:

```
$ cat length-delimited.txt | protoc --encode=StringValue
  wrappers.proto | hexdump -C
00000000  0a 0a 30 31 32 33 34 35  36 37 38
   39                      |..0123456789|
0000000c
```

Here, the first `0a` instance is the field tag + type, the second `0a` instance is the hexadecimal representation of 10, and then we have the ASCII values for each character. To see why 0 turns into 30, you can check the ASCII manual by typing `man ascii` in your terminal and looking for the hexadecimal set. You should have a similar output to the following:

```
30  0     31  1     32  2     33  3     34  4
35  5     36  6     37  7     38  8     39  9
```

Here, the first number of each pair is the hexadecimal value for the second one.

Another kind of message field that will be serialized into a length-delimited type is a repeated field. A repeated field is the equivalent of a list. To write such a field, we simply add the `repeated` keyword before the field type. If we wanted to serialize a list of IDs, we could write the following:

```
repeated uint64 ids = 1;
```

And with this, we could store 0 or more IDs.

Similarly, these fields will be serialized with the length as a prefix. If we take the `ids` field and serialize the numbers from 1 to 9, we will have the following:

```
$ cat repeated.txt | protoc --encode=RepeatedUInt64Values
  wrappers.proto | hexdump -C
00000000  0a 09 01 02 03 04 05 06  07 08 09 |...........|
0000000b
```

This is a list of 9 elements followed by 1, 2, … and so on.

> **Important note**
>
> Repeated fields are only serialized as length-delimited types when they are storing scalar types except for strings and bytes. These repeated fields are considered packed. For complex types or user-defined types (messages), the values will be encoded in a less optimal way. Each value will be encoded separately and prefixed by the type + tag byte(s) instead of having the type + tag serialized only once.

Field tags and wire types

Up until now, you read "tag + type" multiple times and we did not really see what this means. As mentioned, the first byte(s) of every serialized field will be a combination of the field type and the field tag. Let us start by seeing what a field tag is. You surely noticed something different about the syntax of a field. Each time we define a field, we add an equals sign and then an incrementing number. Here's an example:

```
uint64 id = 1;
```

While they look like an assignment of value to the field, they are only here to give a unique identifier to the field. These identifiers, called tags, might look insignificant but they are the most important bit of information for serialization. They are used to tell Protobuf into which field to deserialize which data. As we saw earlier during the presentation of the different serialization algorithms, the field name is not serialized—only the type and the tag are. And thus, when deserialization kicks in, it will see a number and it will know where to redirect the following datum.

Now that we know that these tags are simply identifiers, let us see how these values are encoded. Tags are simply serialized as varints but they are serialized with a wire type. A wire type is a number that is given to a group of types in Protobuf. Here is the list of wire types:

Type	Meaning	Used for
0	Varint	`int32`, `int64`, `uint32`, `uint64`, `sint32`, `sint64`, `bool`, `enum`
1	64-bit	`fixed64`, `sfixed64`, `double`
2	Length-delimited	string, bytes, packed repeated fields
5	32-bit	`fixed32`, `sfixed32`, `float`

Here, 0 is the type for varints, 1 is for 64-bit, and so on.

To combine the tag and the wire type, Protobuf uses a concept called bit packing. This is a technique that is designed to reduce the number of bits into which the data will be serialized. In our case here, the data is the field metadata (the famous tag + type). So, here is how it works. The last 3 bits of the serialized metadata are reserved for the wire type, and the rest is for the tag. If we take the first example that we mentioned in the *Fixed-size numbers* section, where we serialized 42 in a `fixed32` field with tag 1, we had the following:

```
0d 2a 00 00 00
```

This time we are only interested in the `0d` part. This is the metadata of the field. To see how this was serialized, let us turn `0d` into binary (with 0 padding):

```
00001101
```

Here, we have 101 (5) for the wire type—this is the wire type for 32 bits—and we have 00001 (1) for tag 1. Now, since the tag is serialized as a varint, it means that we could have more than 1 byte for that metadata. Here's a reference for knowing the thresholds at which the number of bytes will increase:

Field tag	Size (in bits)
1	5
16	13
2,048	21
262,144	29
33,554,432	37
536,870,911	37

This means that, as fields without values set to them will not be serialized, we need to keep the lowest tags to the fields that are the most often populated. This will lower the overhead needed to store the metadata. In general, 15 tags are enough, but if you come up with a situation where you need more tags, you might consider moving a group of data into a new message with lower tags.

Common types

As of now, if you checked the companion code, you could see that we are defining a lot of "boring" types that are just wrappers around one field. It is important to note that we wrote them by hand to simply give an example of how you would inspect the serialization of certain data. Most of the time, you will be able to use already defined types that are doing the same.

Well-known types

Protobuf itself comes with a bunch of already defined types. We call them *well-known types*. While a lot of them are rarely useful outside of the Protobuf library itself or advanced use cases, some of them are important, and we are going to use some of them in this book.

The ones that we can understand quite easily are the wrappers. We wrote some by hand earlier. They usually start with the name of the type they are wrapping and finish with `Value`. Here is a list of wrappers:

- `BoolValue`
- `BytesValue`
- `DoubleValue`
- `EnumValue`
- `FloatValue`
- `Int32Value`
- `Int64Value`
- `StringValue`
- `UInt32Value`
- `UInt64Value`

These types might be interesting for debugging use cases such as the ones we saw earlier or just to serialize simple data such as a number, a string, and so on.

Then, there are types representing time, such as `Duration` and `Timestamp`. These two types are defined in the exact same way ([Duration | Timestamp] is not proper protobuf syntax, it means that we could replace by either of both terms):

```
message [Duration | Timestamp] {
    // Represents seconds of UTC time since Unix epoch
    // 1970-01-01T00:00:00Z. Must be from 0001-01-
      01T00:00:00Z to
    // 9999-12-31T23:59:59Z inclusive.
```

```
    int64 seconds = 1;

    // Non-negative fractions of a second at nanosecond
    // resolution. Negative
    // second values with fractions must still have non-
    // negative nanos values
    // that count forward in time. Must be from 0 to
    // 999,999,999
    // inclusive.
    int32 nanos = 2;
}
```

However, as their name suggests, they represent different concepts. A `Duration` type is the difference between the start and end time, whereas a `Timestamp` type is a simple point in time.

Finally, one last important well-known type is `FieldMask`. This is a type that represents a set of fields that should be included when serializing another type. To understand this one, it might be better to give an example. Let us say that we have an API endpoint returning an account with `id`, `username`, and `email`. If you wanted to only get the account's email address to prepare a list of people you want to send a promotional email to, you could use a `FieldMask` type to tell Protobuf to only serialize the `email` field. This lets us reduce the additional cost of serialization and deserialization because now we only deal with one field instead of three.

Google common types

On top of well-known types, there are types that are defined by Google. These are defined in the `googleapis/api-common-protos` GitHub repository under the `google/type` directory and are easily usable in Golang code. I encourage you to check all the types, but I want to mention some interesting ones:

- `LatLng`: A latitude/longitude pair storing the values as doubles
- `Money`: An amount of money with its currency as defined by ISO 4217
- `Date`: Year, Month, and Day stored as `int32`

Once again, go to the repository to check all the others. These types are battle-tested and in a lot of cases more optimized than trivial types that we would write. However, be aware that these types might also not be a good fit for your use cases. There is no such thing as a one-size-fits-all solution.

Services

Finally, the last construct that is important to see and that we are going to work with during this book is the service one. In Protobuf, a service is a collection of RPC endpoints that contains two major parts. The first part is the input of the RPC, and the second is the output. So, if we wanted to define a service for our accounts, we could have something like the following:

```
message GetAccountRequest {…}
message GetAccountResponse {…}
service AccountService {
  rpc GetAccount(GetAccountRequest) returns (GetAccountResponse);
  //...
}
```

Here, we define a message representing a request, and another one representing the response and we use these as input and output of our getAccount RPC call. In the next chapter, we are going to cover more advanced usage of the services, but right now what is important to understand is that Protobuf defines the services but does not generate the code for them. Only gRPC will.

Protobuf's services are here to describe a contract, and it is the job of an RPC framework to fulfill that contract on the client and server part. Notice that I wrote *an RPC framework* and not simply gRPC. Any RPC framework could read the information provided by Protobuf's services and generate code out of it. The goal of Protobuf here is to be independent of any language and framework. What the application does with the serialized data is not important to Protobuf.

Finally, these services are the pillars of gRPC. As we are going to see later in this book, we will use them to make requests, and we are going to implement them on the server side to return responses. Using the defined services on the client side will let us feel like we are directly calling a function on the server. If we talk about AccountService, for example, we can make a call to GetAccount by having the following code:

```
res := client.GetAccount(req)
```

Here, client is an instance of a gRPC client, req is an instance of GetAccountRequest, and res is an instance of GetAccountResponse. In this case, it feels a little bit like we are calling GetAccount, which is implemented on the server side. However, this is the doing of gRPC. It will hide all the complex ceremony of serializing and deserializing objects and sending those to the client and server.

Summary

In this chapter, we saw how to write messages and services, and we saw how scalar types are serialized and deserialized. This prepared us for the rest of the book, where we are going to use this knowledge extensively.

In the next chapter, we are going to talk about gRPC, why it uses Protobuf for serialization and deserialization, and what it is doing behind the scenes, and we are going to compare it with REST and GraphQL APIs.

Quiz

1. What is the number 32 representing in the `int32` scalar type?

    A. The number of bits the serialized data will be stored in

    B. The range of numbers that can fit into the scalar type

    C. Whether the type can accept negative numbers or not

2. What is varint encoding doing?

    A. Compressing data in such a way that a smaller number of bytes will be required for serializing data

    B. Turning every negative number into positive numbers

3. What is ZigZag encoding doing?

    A. Compressing data in such a way that a smaller number of bytes will be required for serializing data

    B. Turning every negative number into a positive number

4. In the following code, what is the `= 1` syntax and what is it used for?

    ```
    uint64 ids = 1;
    ```

    A. This is assigning the value 1 to a field

    B. 1 is an identifier that has no other purpose than helping developers

    C. 1 is an identifier that is helping the compiler know into which field to deserialize the binary data.

5. What is a message?

    A. An object that contains fields and represents an entity

    B. A collection of API endpoints

    C. A list of possible states

6. What is an enum?

    A. An object that contains fields and represents an entity

    B. A collection of API endpoints

    C. A list of possible states

7. What is a service?

    A. An object that contains fields and represents an entity

    B. A collection of API endpoints

    C. A list of possible states

Answers

1. B
2. A
3. B
4. C
5. A
6. C
7. B

3

Introduction to gRPC

Now that we have a basic understanding of how data flows over the network and how Protobuf works, we can enter the gRPC world. In this chapter, the goal is to understand what gRPC is doing on top of HTTP/2 and why Protobuf is the perfect fit for gRPC, and also to see that gRPC is a mature technology backed up by major companies in the industry. This will give us a sense of why gRPC is described as "Protobuf over HTTP/2" and make us confident in using it without fearing that the technology is too new and without community.

In this chapter, we're going to cover the following main topics:

- Major use cases for gRPC
- Advantages of using Protobuf
- The role of gRPC on top of Protobuf

Prerequisites

You can find the code for this chapter at `https://github.com/PacktPublishing/gRPC-Go-for-Professionals/tree/main/chapter3`. During this chapter, I will be using protoc to generate Go code out of `.proto` files. This means that you need to make sure you have protoc installed. You can download a zip file from the **Releases** page of the Protobuf GitHub repository (`https://github.com/protocolbuffers/protobuf/releases`); uncompress it and follow the `readme.txt` instructions (note: we do intend to use Well-Known Types in the future so make sure you also install the includes). On top of protoc, you are going to need two protoc plugins: `protoc-gen-go` and `protoc-gen-go-grpc`. The former generates Protobuf code, and the latter generates gRPC code. To add them, you can simply run the following commands:

```
$ go install google.golang.org/protobuf/cmd/protoc-gen-go
  @latest
$ go install google.golang.org/grpc/cmd/protoc-gen-go-
  grpc@latest
```

And finally, make sure that your GOPATH environment variable is in your PATH environment variable. Normally, this is already done for you on the installation of Golang, but if you get any error related to not finding protoc-gen-go or protoc-gen-go-grpc, you will need to do it manually. To get the GOPATH environment variable, you can run the following command:

```
$ go env GOPATH
```

And then, depending on your OS, you can go through the steps of adding the output to your PATH environment variable.

A mature technology

gRPC is not just another new cool framework that you can disregard as being a fad It is a framework that has been battle-tested at scale for over a decade by Google. Originally, the project was for internal use, but in 2016, Google decided to provide an open source version of it that was not tied to the specifics of the company's internal tooling and architecture.

After that, companies such as Uber—and a lot more—migrated their existing services to gRPC for efficiency but also for all the extra features that it offers. Moreover, some open projects such as etcd, which is a distributed key-value store used at the core of Kubernetes, use gRPC for communication across multiple instances.

Recently, Microsoft joined the effort around building a .NET implementation of gRPC. While it is not the goal of this book to explain what it did, it clearly shows an interest in the project. Furthermore, the more that companies such as this are willing to contribute, the more resources will be out there and the greater the community/tooling available will be. The project has a powerful backup, and this is good for all of us.

Now, all of this sounds amazing, but I am aware that most of us will not reach the scale of these giants, so it is important to understand what gRPC is good at. Let us see some use cases where it shines. The first use case that everyone is talking about is communication for microservices. This use case is an appealing one, especially for polyglot microservices. Our job as software engineers is to choose the right job for the right tool and code generation in different languages to enable us to do that.

Another use case is real-time updates. As we saw, gRPC gives us the possibility of streaming data. This comes in multiple flavors such as server streaming, which could be useful for keeping up to date with data such as stock prices. Then, we have client streaming, which could be useful for sensors streaming data to backends. And finally, we have bi-directional streaming, which could be interesting when both the client and server need to be aware of each other's updates, such as messaging apps.

Another important use case is **inter-process communication** (**IPC**). This is communication happening on the same machine between different processes. It can be useful for synchronizing two or more distinct applications, implementing **separation of concerns** (**SOC**) with a modular architecture, or increasing security by having application sandboxing.

Obviously, I presented the most common applications of gRPC that I can see out there, but there are a lot more applications of it, and it is important that you test it on your use case to see whether it fits your requirements. And if you are interested in testing gRPC, you will need to start trying to find out how Protobuf can reduce your payloads and application efficiency.

Why Protobuf?

By now, you should understand that Protobuf provides us with a way of writing data schemas describing how our data should be serialized and deserialized. Then, the Protobuf compiler (protoc) lets us generate some code from these schemas to use the generated types in our code, and these serializable types are exactly what gRPC uses to let user code interact with request and response objects and let the gRPC framework send binary representations of them over the wire.

The binary representation of messages is the biggest reason Protobuf is used as the default data schema for gRPC. The data is serialized in way fewer bytes than traditional data schemas (XML, JSON, and so on). This means that not only can the message be delivered faster, but also that it will be deserialized faster.

> **Important note**
>
> The following experiment is mostly done to show Protobuf's performance. Examples are exaggerated, but this will give you a sense of the additional cost that JSON has during deserialization. The results might vary across the run, OS, and hardware, so if you want to run your own experiment, you can find the benchmarking code and data in the chapter3 folder (https://github.com/PacktPublishing/gRPC-Go-for-Professionals/tree/main/chapter3). To get the data, you will have to unzip it with gzip. You can do that by running the gzip -dk accounts.json.gz or gzip -dk accounts.bin.gz command. After that, to run the experiment, you first need to compile the .proto files with protoc --go_out=proto -Iproto --go_opt=module=https://github.com/PacktPublishing/gRPC-Go-for-Professionals/proto proto/*.proto, and then you can execute the Go code by running go run main.go in the chapter3 folder.

To demonstrate that, we can do a simple experiment—we can generate 100,000 accounts (with an ID + username), run the deserialization 1,000 times, and calculate the mean time needed to deserialize all the data. Here are the results from one of the runs with untrimmed (newlines and spaces) JSON against the Protobuf binary:

```
JSON: 49.773ms
PB: 9.995ms
```

However, most developers do trim their JSON, so here is the result after removing newlines and spaces:

```
JSON: 38.692ms
PB: 9.712ms
```

It is better but still significantly slower than Protobuf.

And finally, we can look at the serialized data size for the data used in the experiment. For the uncompressed JSON versus uncompressed Protobuf, we have the following output:

```
1.3M  accounts.bin
3.1M  accounts.json
```

And for the compressed version (gzip), we have this output:

```
571K accounts.bin.gz
650K accounts.json.gz
```

I encourage you to experiment more with this and especially experiment for your use cases but, unless there is a big mistake in the proto file design, you will find that Protobuf is way more efficient in terms of size and serialization/deserialization time.

On top of providing data serialization, we saw that Protobuf also has a concept of service, which is a contract between clients and servers. While this concept is not linked exclusively to gRPC (you could generate code wrapping other frameworks), gRPC uses that to generate the appropriate API endpoints. This provides us with type safety on both the client and server sides. If we try to send incorrect data and we work in a compiled language, we will get a compilation error instead of getting an error at runtime. This dramatically shortens the feedback loop for developers and reduces the area of possible failures in our code.

Finally, Protobuf itself is language-agnostic. This means that this is an independent data schema, and it can be shared across multiple projects. If you have code written in C++ for some microcontroller, sending data to a backend written in Go, which is in turn sending data to a web frontend in JS, you can simply share the same Protobuf file and generate your models with protoc. You do not have to rewrite them every time in the different projects. This decreases the area that needs to be updated on adding or updating features and provides an interface that multiple teams need to agree on.

In the end, Protobuf enables faster communication (for example, through gRPC) by creating smaller payloads, it provides us with type safety on both ends of the communication, and it does all that across multiple languages so that we can use the right tool for the right job.

What is gRPC doing?

gRPC is described as "Protobuf over HTTP/2." This means that gRPC will generate all the communication code wrapping the gRPC framework and stand on Protobuf's shoulders to serialize and deserialize data. To know which API endpoints are available on the client and server, gRPC will look at the services defined in our .proto files, and from that, it will learn the basic information needed to generate some metadata and the functions needed.

The first thing to understand with gRPC is that there are multiple implementations. In Go, for example, you get a pure implementation of gRPC. This means that the entire code generation process and communication is written in Go. Other languages might have similar implementations, but a lot of them are wrappers around the C implementation. While we do not need to know anything about them in the context of this book, it is important to know that they are available because it explains the presence of plugins for the protoc compiler.

As you know, there are a lot of languages out there. Some are relatively new, and some are pretty old, so staying on top of every language's evolution is practically infeasible. That is why we have protoc plugins. Every developer or company interested in supporting a language can write such a plugin to generate code that will send Protobuf over HTTP/2. This is, for example, the case for Swift support, which was added by Apple.

Since we are talking about Go, we want to look at what kind of code is generated to get a sense of how gRPC works but also to know how to debug and where to look for function signatures. Let us start with a simple service—in `proto/account.proto`, we have the following:

```
syntax = "proto3";

option go_package = "github.com/PacktPublishing/
  gRPC-Go-for-Professionals";

message Account {
  uint64 id = 1;
  string username = 2;
}

message LogoutRequest {
  Account account = 1;
}

message LogoutResponse {}

service AccountService {
  rpc Logout (LogoutRequest) returns (LogoutResponse);
}
```

In this service, we have an API endpoint named Logout that takes as a parameter LogoutRequest (a wrapper around Account) and returns a LogoutResponse parameter. LogoutResponse is an empty message because we want to send the account for which the session needs to be stopped and we do not need any result, just an indicator that the call went well.

Then, to generate Protobuf and gRPC code out of this, we will run the following command:

```
$ protoc --go_out=. \
         --go_opt=module=github.com/PacktPublishing/gRPC-Go-for-
           Professionals \
         --go-grpc_out=. \
         --go-grpc_opt=module=github.com/PacktPublishing/gRPC-Go-for-
           Professionals \
         proto/account.proto
```

We already saw with Protobuf that the messages will be turned into structs, but now we also have a _grpc.pb.go file that contains the gRPC communication code.

The server

Let us look at what was generated on the server side first. We are going to start from the bottom of the file with the service descriptor. But first, we need to know what a descriptor is. In Protobuf and gRPC context, a descriptor is a meta object that represents Protobuf code. This means that, in our case, we have a Go object representing a service or other concepts. In fact, we did not deep-dive into it in the previous chapter but if you look at the generated code for Account, you will also find Desc being mentioned.

For our AccountService service, we have the following descriptor:

```
var AccountService_ServiceDesc = grpc.ServiceDesc{
  ServiceName: "AccountService",
  HandlerType: (*AccountServiceServer)(nil),
  Methods: []grpc.MethodDesc{
    {
      MethodName: "Logout",
      Handler: _AccountService_Logout_Handler,
    },
  },
  Streams: []grpc.StreamDesc{},
  Metadata: "account.proto",
}
```

This means that we have a service called AccountService that is linked to a type called AccountServiceServer, and this service has a method called Logout that should be handled by a function called _AccountService_Logout_Handler.

You should find the handler above the service descriptor. This looks like the following (simplified):

```
func _AccountService_Logout_Handler(srv interface{}, ctx
  context.Context, dec func(interface{}) error, interceptor
```

```
  grpc.UnaryServerInterceptor) (interface{}, error) {
in := new(LogoutRequest)
if err := dec(in); err != nil {
  return nil, err
}
if interceptor == nil {
  return srv.(AccountServiceServer).Logout(ctx, in)
}
//...
}
```

This handler is responsible for creating a new object of type `LogoutRequest` and populating it before passing it to the `Logout` function in an object of type `AccountServiceServer`. Note here that we are going to assume that we always have an interceptor equal to `nil` because this is a more advanced feature, but later, we are going to see an example of how to set one up and use it.

Finally, we see the `AccountServiceServer` type being mentioned. Here is what it looks like:

```
type AccountServiceServer interface {
  Logout(context.Context, *LogoutRequest) (*LogoutResponse,
    error)
  mustEmbedUnimplementedAccountServiceServer()
}
```

This is a type that contains the function signatures of our RPC endpoints and a `mustEmbed UnimplementedAccountServiceServer` function.

Before going to the `Logout` function, let us understand `mustEmbedUnimplemented AccountServiceServer`. This is an important concept for gRPC because it is here to provide a forward-compatible implementation of our service, and what it means is that older versions of our API will be able to communicate with newer ones without crashing.

If you check under the definition of `AccountServiceServer`, you will see the following:

```
// UnimplementedAccountServiceServer must be embedded to
  have forward compatible implementations.
type UnimplementedAccountServiceServer struct {
}

func (UnimplementedAccountServiceServer)
  Logout(context.Context, *LogoutRequest) (*LogoutResponse,
    error) {
  return nil, status.Errorf(codes.Unimplemented, "method
    Logout not implemented")
}
```

With that, we can understand that this UnimplementedAccountServiceServer type must be embedded somewhere, and this *somewhere* is in a type that we are going to define later in this book when we are going to write our API endpoints. We are going to have the following code:

```
type struct Server {
  UnimplementedAccountServiceServer
}
```

This is called type embedding, and this is the way Go goes about adding properties and methods from another type. You might have heard the advice to prefer composition over inheritance, and that is just that. We add the methods' definitions from UnimplementedAccountServiceServer to Server. This will let us have the default implementations that return method Logout not implemented generated for us. This means that if a server without a full implementation receives a call on one of its unimplemented API endpoints, it will return an error but not crash because of the non-existent endpoint.

Once we understand that, the Logout method signature is trivial. As mentioned, later we are going to define our own server type that embeds the UnimplementedAccountServiceServer type, and we are going to override the Logout function with the implementation. Any call to Logout will then be redirected to the implementation and not to the default generated code.

The client

The generated code for the client is even simpler than the server code. We have an interface called AccountServiceClient that contains all the API endpoints:

```
type AccountServiceClient interface {
  Logout(ctx context.Context, in *LogoutRequest, opts
    ...grpc.CallOption) (*LogoutResponse, error)
}
```

And we have the actual implementation of that interface, called accountServiceClient:

```
type accountServiceClient struct {
  cc grpc.ClientConnInterface
}

func NewAccountServiceClient(cc grpc.ClientConnInterface)
  AccountServiceClient {
  return &accountServiceClient{cc}
}

func (c *accountServiceClient) Logout(ctx context.Context,
```

```
  in *LogoutRequest, opts ...grpc.CallOption)
    (*LogoutResponse, error) {
  out := new(LogoutResponse)
  err := c.cc.Invoke(ctx, "/AccountService/Logout", in,
    out, opts...)
  if err != nil {
    return nil, err
  }
  return out, nil
}
```

We can notice one important thing in this piece of code. We have an endpoint route called /AccountService/Logout. If you take a look back at the AccountService_ServiceDesc variable described in the section titled *The server*, you will find out that this route is a concatenation of the ServiceName and MethodName properties. This will let the server know how to route that request to the _AccountService_Logout_Handler handler

That is all. We can see that gRPC is handling all the boilerplate code to call an endpoint. We just need to create an object following the AccountServiceClient interface by calling NewAccountServiceClient, and then with that object, we can call the Logout member.

The read/write flow

Now that we have seen what Protobuf and gRPC are, it is time to go back to the read/write flow that we presented in *Chapter 1*. The goal of doing this is to make it a little bit more detailed and include what we learned.

As a quick reminder, we saw that they are mostly three levels when writing and reading data. We have the user code, the gRPC framework, and the transport layers. What is interesting for us here is mostly the user code. We did not go into too much detail in *Chapter 1* but now that we are equipped with more knowledge on what gRPC is doing, we can understand the process more clearly.

The user-code layer is the code that developers write and interacts with the gRPC framework. For the client, this is calling the endpoints, and for the server, this is the implementation of the endpoints. If we keep going with our AccountService service, we can give a concrete example of the read/write flow.

The first thing that we can do is separate the user-code layer into two parts: the implementation and the generated code. Furthermore, in *Chapter 1*, we gave a rather generic schema where we described the overall flow and drew a cryptic component called Other Actor. Let us now split the server and the client into two different actors, and we have the following system:

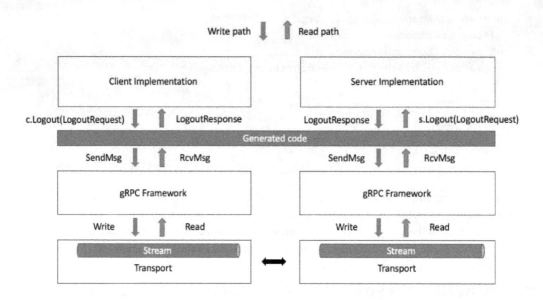

Figure 3.1 – Specialization of the read/write flow for AccountService

> **Important note**
>
> In the preceding diagram, I am using abbreviations "c" and "s" to refer to the client and server respectively. "c" is an instance of `AccountServiceClient` created by `NewAccountServiceClient`, and "s" is an instance of a type defined in `Implementation` that defines the `Logout` function.

We can see a few important things happening once we expand the diagram. The first interesting concept is that the generated code is shared across the different communication actors. We saw that the gRPC Go plugin will generate a single file containing the server and client types. This means that this file should be shared between all actors written in Go.

We can also notice that the gRPC framework and generated code abstract everything for us. This lets us focus only on calling an endpoint with a `Request` object and on writing the endpoint handling that `Request` object and returning a `Response` object. This highly limits the amount of code we need to write and thus makes our code more testable because we need to focus on less code.

Finally, the last important thing to notice is that we can limit ourselves to reading the generated code to understand the parameters and return types of each of our endpoints. This is helpful because either the generated code will be picked up by your IDE and you will have autocompletion or you can simply check one file to get the information you need.

Why does gRPC matter?

Now that we have a sense of what gRPC is, we can get into why it matters. To explain gRPC's role, we are going to compare it with two other ways of performing client/server communication. The first one is the traditional REST API architecture based on HTTP and JSON, and the second one is GraphQL.

REST

While I am assuming that most of you reading this book are familiar with REST APIs, I still believe that it is important to introduce the principles of designing such APIs. It will help us understand in which ways gRPC is like REST and in which it differs.

A REST API, as with every other technology in this comparison study, is an interface between an information provider and a consumer. When writing such an API, we expose endpoints on specific URLs (routes) that can be used by a client to create, read, update, and delete resource(s).

However, REST APIs are different from gRPC and GraphQL. The main difference is that REST is not a framework—it is a set of architectural practices that can be implemented in different ways. The main constraints are the following:

- The client, server, and resources are the main entities during communication. The client requests resources from the server and the server returns the relevant resources.

- Requests and responses are managed by HTTP. `GET` is used to read resources, `POST` to create resources, `PUT` to update resources, `PATCH` to update part of a resource, and `DELETE` to remove resources.

- No client-related information should be stored between requests. This is a stateless communication, and each request is separate.

Finally, even though such APIs are not bound to any data format, the most used one is JSON. This is mostly because JSON has a wide community, and a lot of languages and frameworks can handle this data format.

GraphQL

GraphQL is presented as a query language for APIs. It lets developers write data schemas that describe the data available, and it lets them query a specific set of fields present in the schema.

As it lets us write queries, we can have more generic endpoints and request only the fields that we are interested in for a certain feature. This solves the problem of overfetching and underfetching because we only get the amount of data that we asked for.

On top of all of this, as GraphQL mainly uses JSON as data format and uses explicit types and comments in its own data schema, this makes GraphQL more readable and self-documenting. This makes this technology more mature for companies at scale because we can do type-checking at compile time and shorten the feedback loop, and we do not have to separate documentation and code, thus it will be less likely to be not in sync.

Comparison with gRPC

Now that we have had an overview of what each technology does, we can get started with comparing them with gRPC. We are going to focus on the biggest differentiators between these four ways of designing an API. These differentiators are the following:

- The transport, data format, and data schema used for communication

- Separation of concern of API endpoints

- The developers' workflow when writing APIs

- The convenience of out-of-the-box features

Transport, data format, and data schema

In this regard, GraphQL and REST APIs are similar. They both use HTTP/1.1 for the underlying transport and, more often than not, developers use JSON for sending structured data. On the other side, gRPC uses HTTP/2 and Protobuf by default. This means that, with gRPC, we have smaller payloads to send over the wire and we have more efficient connection handling.

There are certain things to be more careful about when we are dealing with Protobuf than when we have to deal with JSON. Protobuf provides implicit default values depending on the type of field, and these default values do not get serialized into the final binary. For example, `int32` has a default value of 0. This means that we cannot differentiate between the value 0 being set or whether the field was not set. Of course, there are ways of dealing with that, but it makes the client-side usage a little bit more involved. In this respect, GraphQL handles default values differently. We can pass the default values as parameters of our endpoints, and this means that we can handle particular cases in a more user-friendly way.

Finally, it is important to mention that all these technologies are quite flexible regarding the kind of data format that you can transport over the wire. REST APIs handle binary and other kinds of data, GraphQL can also accept binary data, and gRPC can send JSON data. However, problems come with this flexibility. If you are using binary over a REST API, you let both the client and the server interpret what this binary means. There is no type safety and we need to handle serialization/deserialization errors that otherwise would be handled by libraries or frameworks. If you use binary with GraphQL, you are greatly reducing the number of community tools that you can use. And finally, if you use JSON with gRPC, you are losing all the advantages of Protobuf.

Separation of concern of API endpoints

Designing separation of concern for APIs can be tricky and lead to problems such as underfetching or overfetching. GraphQL was designed to solve these problems of getting too much or too little data when making a request. With it, you can simply ask for a specific set of fields that you need for a certain feature. While it is possible to do a similar thing with gRPC and REST APIs, it remains non-user-friendly when your API is facing external users.

However, separation of concern in APIs can help us with a few things. First, it can help us reduce the scope of testing for an endpoint. Instead of thinking about all the possible inputs and outputs that an endpoint might have, we are only focusing on a specific input and a specific output.

And second, having smaller and more specific endpoints will help in the case of API abuse. As we can clearly know which request was made to which endpoint, we can rate-limit them per client and thus secure our APIs. With more flexible API endpoints such as in GraphQL, this is intrinsically harder to implement because we need to ponder whether to rate limit on the whole route, a specific input, or just a simple query.

The developers' workflow

Another important aspect of these technologies that is often overlooked is the developers' workflow when writing an API. With REST APIs, we are mostly working on the server and the client separately, and this process is error-prone. If we do not have specifications on what data to expect, we are in for long sessions of debugging. Furthermore, even if we have specifications on the data, developers are humans and humans make mistakes. The client might have expected a certain kind of data, but the server is sending another.

Now, this is not a problem that concerns only REST APIs—gRPC and GraphQL APIs also have this problem. However, the problem scope is reduced because we can make sure that only a certain type can be used as a request and another as a response. This lets us focus on the happy path instead of writing code that is checking whether the serialization and deserialization failed.

The gRPC and GraphQL way of developing APIs is called **schema-driven development (SDD)**. We first focus on writing a schema that defines all higher-level requirements of our API, and then we dive into the implementation. This greatly reduces the scope of errors that we can get at runtime and also shortens the feedback loop for developers. We cannot send a string instead of an int, and it will tell us that at compile time. This also makes the scope of tests much smaller because we can now focus on the feature itself and not on the many possible errors that could happen due to external problems.

Convenience

Finally, another overlooked topic is how convenient using technology is. This can be due to the community developing tools or simply out-of-the-box features coming with the framework. In this case, technologies using JSON often have more tools and support. This is the case because JSON has been widely used for a long time and it is attractive because of its human readability.

However, even with the lack of tool compared to JSON-backed APIs, gRPC was designed on principles that helped Google scale and secure its products, it has a lot of amazing features that you can get without any extra dependencies. gRPC has interceptors, TLS authentication, and many other high-end features built in as part of the official framework, and thus it is simpler to write secure and performant code.

Finally, GraphQL is the only technology of the three that is explicit about endpoints having side effects. This can be documented for gRPC or REST APIs; however, this cannot be checked statically. This is important because this makes the APIs' users more aware of what is happening in the background, and it might lead to better choices for appropriate routes.

Summary

To summarize, gRPC is a mature technology adopted by tech giants but also the open source community to create efficient and performant client/server communication. This is not only true in the distributed system but also in the local environment with the use of IPC. gRPC uses Protobuf by default due to its compact binary serialization and fast deserialization but also for its type safety and language agnosticism. On top of that, gRPC generates code to send Protobuf over HTTP/2. It generates a server and a client for us so that we do not have to think about the details of communication. All the details are handled by the gRPC framework.

In the next chapter, we are finally going to get our hands dirty. We are going to set up a gRPC project, make sure that our code generation is working properly, and write some boilerplate code for both the server and the client.

Quiz

1. What is one of the reasons Protobuf is the default data format for gRPC?

    A. The serialized data is human-readable

    B. It is dynamically typed

    C. It is type-safe

2. In the Go implementation, which component generates server/client code?

    A. Protoc

    B. The gRPC Go plugin

    C. Other

3.   What are service descriptors in the context of gRPC-generated code?

    A.   They describe which endpoints a service has and how to handle requests

    B.   They describe how to return responses

    C.   Both of these

4.   How will the user code be able to implement the logout endpoint?

    A.   By writing code in the generated `Logout` function

    B.   By creating a copy of the generated code and editing it

    C.   By using type embedding with the generated server and implementing `Logout` for that type

Answers

1.   C

2.   B

3.   A

4.   C

4

Setting Up a Project

As the chapter title suggests, we are going to set up a gRPC project from scratch. We are first going to create our Protobuf schema as we are doing schema-driven development. Once the schema is created, we will generate Go code. Finally, we are going to write the templates for the server and client so that we can reuse them later in this book.

In this chapter, we're going to cover the following main topics:

- Common gRPC project architecture
- Generating Go code out of a schema
- Writing reusable server/client templates

Prerequisites

I assume that you already have protoc installed from the last chapter. If you do not, this is the right time to install it because without it, you will not benefit as much from this chapter.

In this chapter, I will show common ways of setting up a gRPC project. I will use protoc, Buf, and Bazel. Thus, depending on the one(s) you are interested in, you will have to download the tool(s). Buf is an abstraction over protoc that lets us run protoc commands more easily. On top of that, it provides features such as linting and detecting breaking changes. You can download Buf from here: `https://docs.buf.build/installation`. I will also use Bazel to automatically generate Go code from Protobuf and the binary of our server and client. If you are interested in using it, you can check the installation documentation (`https://github.com/bazelbuild/bazelisk#installation`).

Finally, you can find the code for this chapter under the `chapter4` folder in the accompanying GitHub repository (`https://github.com/PacktPublishing/gRPC-Go-for-Professionals/tree/main/chapter4`).

Creating a .proto file definition

Since the goal of this chapter is to write a template that we can use for later projects, we are going to create a dummy proto file that will let us test whether our build system is working properly or not. This dummy proto file will contain both a message and a service because we want to test code generation for both Protobuf and gRPC.

The message, called `DummyMessage`, will be defined as follows:

```
message DummyMessage {}
```

The service, called `DummyService`, will be defined as follows:

```
service DummyService {}
```

Now, because we are planning to generate Golang code, we still need to define an option called `go_package` and set its value to the name of the Go module concatenated with the name of the subfolder containing the proto files. This option is important because it lets us define the package in which the generated code should be. In our case, the project architecture is the following:

```
.
├── client
│   └── go.mod
├── go.work
├── proto
│   ├── dummy
│   │   └── v1
│   │       └── dummy.proto
│   └── go.mod
└── server
    └── go.mod
```

We have a monorepo (Go workspace) with three submodules: `client`, `proto`, and `server`. We create each submodule by going into each folder and running the following command:

```
$ go mod init github.com/github.com/PacktPublishing/
gRPC-Go-for-Professionals/
$FOLDER_NAME
```

`$FOLDER_NAME` should be replaced with the name of the folder you are currently in (`client`, `proto`, or `server`).

To make the process a little bit quicker, we can create a command that will list the folder in the `root` directory and execute the `go` command. To do so, you can use the following UNIX (Linux/macOS) command:

```
$ find . -maxdepth 1 -type d -not -path . -execdir sh -c "pushd {}; go
mod init 'github.com/PacktPublishing/
gRPC-Go-for-Professionals/{}';
popd" ";"
```

If you are on Windows, you can use PowerShell to run the following:

```
$ Get-ChildItem . -Name -Directory | ForEach-Object { Push-
Location $_; go mod init "github.com/PacktPublishing/
gRPC-Go-for-Professionals/$_" ;
Pop-Location }
```

After that, we can create the workspace file. We do so by going to the root of the project (`chapter4`) and running the following:

```
$ go work init client server proto
```

We now have the following `go.work`:

```
go 1.20

use (
  ./client
  ./proto
  ./server
)
```

Each submodule has the following `go.mod`:

```
module $MODULE_NAME/$SUBMODULE_NAME

go 1.20
```

Here, we'll replace $MODULE_NAME with a URL such as `github.com/PacktPublishing/gRPC-Go-for-Professionals` and $SUBMODULE_NAME, with the respective name of the folder containing the file. In the case of `go.mod` in the client, we will have the following:

```
module github.com/PacktPublishing/gRPC-Go-for-Professionals/client

go 1.20
```

Finally, we can complete `dummy.proto` by adding the following line to our file:

```
option go_package = "github.com/PacktPublishing/
gRPC-Go-for-Professionals/
proto/dummy/v1";
```

We now have the following `dummy.proto`:

```
syntax = "proto3";

option go_package = "github.com/PacktPublishing/
gRPC-Go-for-Professionals/proto/dummy/v1";

message DummyMessage {}
service DummyService {}
```

That is all we need to test the code generation out of our `dummy.proto`.

Generating Go code

To stay impartial in terms of the tools you need to generate code, I will present three different tools from the lowest level to the highest. We are going to start by seeing how to manually generate code with protoc. Then, because we do not want to write lengthy command lines all the time, we are going to see how to make this generation easier with Buf. Finally, we are going to see how to use Bazel to integrate the code generation as part of our build.

> **Important note**
>
> In this section, I'm going to show basic ways of compiling your proto files. Most of the time, these commands will get you by, but sometimes you might have to check each tool's documentation. For protoc, you can run `protoc --help` and get a list of the options. For Buf, you can go to the online documentation: `https://docs.buf.build/installation`. For Bazel, you also have online documentation at `https://bazel.build/reference/be/protocol-buffer`.

Protoc

Using protoc is the manual way of generating code out of proto files. This technique might be fine if you are only dealing with a few proto files and you do not have a lot of dependencies between files (imports). Otherwise, this will be quite painful, as we will see.

However, I still believe that we should learn the protoc command a bit so that we get a sense of what we can do with it. Also, higher-level tools are based on protoc, so this will help us understand its different features.

With our `dummy.proto` file from the previous section, we can run protoc in the root folder (`chapter4`) like the following:

```
$ protoc --go_out=. \
         --go_opt=module=github.com/PacktPublishing/
             gRPC-Go-for-Professionals \
         --go-grpc_out=. \
         --go-grpc_opt=module=github.com/PacktPublishing/
             gRPC-Go-for-Professionals \
         proto/dummy/v1/dummy.proto
```

Now, this might look a bit scary and, in fact, is not the shortest command that you could write to do this. I am going to show you a more compact one when we talk about Buf. But first, let us dissect the preceding command into parts.

Before discussing `--go_out` and `--go-grpc_out`, let us look at `--go_opt=module` and `--go-grpc_opt=module`. These options are telling protoc about the common module to be stripped out by the value passed for the `go_package` option in our `proto` file. Say we have the following:

```
option go_package = "github.com/PacktPublishing/
gRPC-Go-for-Professionals/proto/dummy/v1";
```

Then, `--go_opt=module=github.com/PacktPublishing/gRPC-Go-for-Professionals` will strip the value after `module=` from our `go_package`, so now we only have `/proto/dummy/v1`.

Now that we understand this, we can get to `--go_out` and `--go-grpc_out`. These two options tell protoc where to generate the Go code. In our case, it seems that we are telling protoc to generate our code at the root level, but in fact, because it is combined with the two previous options, it will generate the code right next to the proto file. This is due to the stripping of the package, which leads protoc to generate the code in the folder `/proto/dummy/v1` package.

Now, you can see how painful it might be to write that kind of command all the time. Most people do not do that. They either write a script to do it automatically or use other tools, such as Buf.

Buf

For Buf, we have a little bit more setup to do to generate code. At the root of the project (`chapter4`), we are going to create a Buf module. To do that, we can simply run the following:

```
$ buf mod init
```

This creates a file called buf.yaml. This file is where you can set project-level options such as linting or even tracking breaking changes. These are beyond the scope of this book, but if you are interested in this tool, check out the documentation (https://buf.build/docs/tutorials/getting-started-with-buf-cli/).

Once we have that, we need to write the configuration for generation. In a file called buf.gen.yaml, we will have the following:

```
version: v1
plugins:
  - plugin: go
    out: proto
    opt: paths=source_relative
  - plugin: go-grpc
    out: proto
    opt: paths=source_relative
```

Here, we are defining the use of the Go plugin for Protobuf and gRPC. For each, we are saying that we want to generate the code in the proto directory, and we are using another --go_opt and --go-grpc_opt for protoc, which is paths=source_relative. When this is set, the generated code is placed in the same directory as the input file (dummy.proto). So, in the end, Buf is running something like what we did in the protoc section. It is running the following:

```
$ protoc --go_out=. \
         --go_opt=paths=source_relative \
         --go-grpc_out=. \
         --go-grpc_opt=paths=source_relative \
         proto/dummy/v1/dummy.proto
```

In order to run generation using Buf, we simply need to run the following command (in chapter4):

```
$ buf generate proto
```

Using Buf is pretty common for mid-size or large projects. It helps with automation of code generation and is easy to get started with. However, you might have noticed that you need to generate the code in one step and then build your Go application. Bazel will help us consolidate everything in one step.

Bazel

> **Important note**
>
> In this section, I will be using variables called GO_VERSION, RULES_GO_VERSION, RULES_GO_SHA256, GAZELLE_VERSION, GAZELLE_SHA256, and PROTO_VERSION. We have not included these variables in this section to ensure that the book remains easily updated. You can find the versions in the versions.bzl file in the chapter4 folder (https://github.com/PacktPublishing/gRPC-Go-for-Professionals/tree/main/chapter4).

Bazel is a little bit trickier to set up, but it is worth the effort. Once you have your build system up and running, you will be able to build the whole application (generation and build) and/or run it in one command.

In Bazel, we start by defining a file called WORKSPACE.bazel at the root level. In this file, we define all the dependencies for our project. In our case, we have dependencies on Protobuf and Go. On top of that, we are also going to add a dependency to Gazelle, which will help us create the BUILD.bazel file needed to generate our code.

So, in WORKSPACE.bazel, before anything else, we are going to define our workspace name, import our version variables, and import some utilities to clone Git repositories and download archives:

```
workspace(name = "github_com_packtpublishing_grpc_go_for_
professionals")

load("//:versions.bzl",
  "GO_VERSION",
  "RULES_GO_VERSION",
  "RULES_GO_SHA256",
  "GAZELLE_VERSION",
  "GAZELLE_SHA256",
  "PROTO_VERSION"
)
load("@bazel_tools//tools/build_defs/repo:http.bzl",
  "http_archive")
load("@bazel_tools//tools/build_defs/repo:git.bzl",
  "git_repository")
```

After that, we are going to define the dependency for Gazelle:

```
http_archive(
  name = "bazel_gazelle",
  sha256 = GAZELLE_SHA256,
  urls = [
    "https://mirror.bazel.build/github.com/bazelbuild/
    bazel-gazelle/releases/download/%s/bazel-gazelle-
    %s.tar.gz" % (GAZELLE_VERSION, GAZELLE_VERSION),
    "https://github.com/bazelbuild/bazel-gazelle/
    releases/download/%s/bazel-gazelle-%s.tar.gz" %
    (GAZELLE_VERSION, GAZELLE_VERSION),
  ],
)
```

Then, we need to pull the dependency for building Go binaries, applications, and so on:

```
http_archive(
  name = "io_bazel_rules_go",
  sha256 = RULES_GO_SHA256,
  urls = [
    "https://mirror.bazel.build/github.com/bazelbuild/
    rules_go/releases/download/%s/rules_go-%s.zip" %
    (RULES_GO_VERSION, RULES_GO_VERSION),
    "https://github.com/bazelbuild/rules_go/releases/
    download/%s/rules_go-%s.zip" % (RULES_GO_VERSION,
    RULES_GO_VERSION),
  ],
)
```

Now that we have that, we can pull the dependencies of `rules_go`, set the toolchain for building the Go project, and tell Gazelle where to find our `WORKSPACE.bazel` file:

```
load("@io_bazel_rules_go//go:deps.bzl",
  "go_register_toolchains", "go_rules_dependencies")
load("@bazel_gazelle//:deps.bzl", "gazelle_dependencies")

go_rules_dependencies()

go_register_toolchains(version = GO_VERSION)

gazelle_dependencies(go_repository_default_config =
  "//:WORKSPACE.bazel")
```

Finally, we pull the dependency for Protobuf and load its dependencies:

```
git_repository(
  name = "com_google_protobuf",
  tag = PROTO_VERSION,
  remote = "https://github.com/protocolbuffers/protobuf"
)

load("@com_google_protobuf//:protobuf_deps.bzl",
  "protobuf_deps")

protobuf_deps()
```

Our `WORKSPACE.bazel` is complete.

Let us now move to `BUILD.bazel` at the root level. In this file, we are going to define the command to run Gazelle, and we are going to let Gazelle know the Go module name and that we want it to not consider the Go files in the `proto` directory. We do so because otherwise, Gazelle would think that the Go file in the `proto` directory should also have its own Bazel target file and this might create problems down the road:

```
load("@bazel_gazelle//:def.bzl", "gazelle")

# gazelle:exclude proto/**/*.go
# gazelle:prefix github.com/PacktPublishing/gRPC-Go-for-Professionals
gazelle(name = "gazelle")
```

> **Note**
>
> With that, we can now run the following:
>
> **$ bazel run //:gazelle**
>
> If you have installed Bazel through Bazelisk, Bazel will try to get its newest version each time you run a bazel command. In order to avoid this, you can create a file called .bazelversion containing the version of bazel you currently have installed. You can find the version by typing
>
> **bazel --version**
>
> An example can be found in the chapter4 folder.

After the dependencies are pulled up and compiled, you should be able to see a `BUILD.bazel` file generated in the `proto/dummy/v1` directory. The most important part of this file is the following `go_library`:

```
go_library(
  name = "dummy",
  embed = [":v1_go_proto"],
  importpath = "github.com/PacktPublishing/gRPC-Go-for-
    Professionals/proto/dummy/v1",
  visibility = ["//visibility:public"],
)
```

Later, we will use this library and link it to our binary. It contains all the generated code that we need to get started.

Server boilerplate

Our build system is ready. We can now focus on the code. But before going into that, let us define what we want. In this section, we want to build a template for the gRPC server that we can reuse for later chapters and even later projects outside of the book. To do that, there are a few things that we want to avoid:

- Implementation details such as service implementation
- Specific connection options
- Setting an IP address as constant

We can solve these by not caring about the generated code anymore. It was just for testing our build system. Then, we will default to an insecure connection for testing. Finally, we will take the IP address as an argument for our program.

Let us do that step by step:

1. We are first going to need to add the gRPC dependency to `server/go.mod`. So, in the `server` directory, we can type the following command:

    ```
    $ go get google.golang.org/grpc
    ```

2. Then, we are going to take the first argument passed to the program and return a usage message if no argument is passed:

    ```
    args := os.Args[1:]

    if len(args) == 0 {
      log.Fatalln("usage: server [IP_ADDR]")
    }

    addr := args[0]
    ```

3. After that, we need to listen for the incoming connection. We can do that with the `net.Listen` provided in Go. This listener will need to be closed at the end of the program. This might be when a user kills it or if the server fails. And obviously, if we get an error during the construction of that listener, we just want the program to fail and let the user know:

    ```
    lis, err := net.Listen("tcp", addr)

    if err != nil {
      log.Fatalf("failed to listen: %v\n", err)
    }

    defer func(lis net.Listener) {
      if err := lis.Close(); err != nil {
        log.Fatalf("unexpected error: %v", err)
      }
    }(lis)
    log.Printf("listening at %s\n", addr)
    ```

4. Now, with all of this, we can start creating a `grpc.Server`. We first need to define some connection options. As this is a template for future projects, we are going to keep the options empty. With this array of `grpc.ServerOption` objects, we can create a new gRPC server. This is the server that we will use later to register endpoints. After that, we will need to close the server at some point, so we use a `defer` statement for that. Finally, we call a function called `Serve` on the `grpc.Server` that we created. This takes the listener as a parameter. It can fail, so if there is an error, we return that to the client:

```
opts := []grpc.ServerOption{}
s := grpc.NewServer(opts...)
//registration of endpoints

defer s.Stop()
if err := s.Serve(lis); err != nil {
  log.Fatalf("failed to serve: %v\n", err)
}
```

In the end, we have the following `main` function (`server/main.go`):

```
package main

import (
  "log"
  "net"
  "os"

  "google.golang.org/grpc"
)

func main() {
  args := os.Args[1:]

  if len(args) == 0 {
    log.Fatalln("usage: server [IP_ADDR]")
  }

  addr := args[0]
  lis, err := net.Listen("tcp", addr)

  if err != nil {
    log.Fatalf("failed to listen: %v\n", err)
  }

  defer func(lis net.Listener) {
    if err := lis.Close(); err != nil {
```

```
            log.Fatalf("unexpected error: %v", err)
        }
    }(lis)

    log.Printf("listening at %s\n", addr)

    opts := []grpc.ServerOption{}
    s := grpc.NewServer(opts...)

    //registration of endpoints

    defer s.Stop()
    if err := s.Serve(lis); err != nil {
        log.Fatalf("failed to serve: %v\n", err)
    }
}
```

We are now able to run our server by running the go run command on our server/main.go. We can terminate the execution by using *Ctrl + C*:

```
$ go run server/main.go 0.0.0.0:50051
listening at 0.0.0.0:50051
```

Bazel

If you want to use Bazel, you will need a couple of extra steps. The first step is to update the BUILD. bazel in our root directory. In there, we are going to use a Gazelle command that will detect all the dependencies needed for our project and dump them in a file called deps.bzl. So, after the gazelle command, we can just add the following:

```
gazelle(
    name = "gazelle-update-repos",
    args = [
        "-from_file=go.work",
        "-to_macro=deps.bzl%go_dependencies",
        "-prune",
    ],
    command = "update-repos",
)
```

Now, we can run the following:

```
$ bazel run //:gazelle-update-repos
```

After it has finished detecting all the dependencies of the `server` module, it will create a `deps.bzl` file and link it inside our `WORKSPACE.bazel`. You should have the following line in the workspace file:

```
load("//:deps.bzl", "go_dependencies")

# gazelle:repository_macro deps.bzl%go_dependencies
go_dependencies()
```

Finally, we can rerun our `gazelle` command to make sure it creates the `BUILD.bazel` file for our server. We run the following:

```
$ bazel run //:gazelle
```

We then get our `BUILD.bazel` in the `server` directory. The most important thing to note is that in this file, we can see Bazel linked gRPC to `server_lib`. We should have something like this:

```
go_library(
  name = "server_lib",
  srcs = ["main.go"],
  deps = [
    "@org_golang_google_grpc//:go_default_library",
  ],
  #...
)
```

We can now run our server in an equivalent way as the `go run` command:

> **Note**
>
> This command will pull Protobuf and build it. Recent Protobuf version require to be built with C++14 or newer. You can tell bazel to automatically specify the C++ version to build Protobuf with in a file called .bazelrc. In order to keep this chapter version-independent, we recommend you to check the .bazelrc file in the chapter4 of the Github repository. You can copy paste the file in your project folder.

```
$ bazel run //server:server 0.0.0.0:50051
listening at 0.0.0.0:50051
```

And so, we are done with the server. This is a simple template that will let us create new servers pretty easily in the next chapters. It is listening on a given port and waiting for some requests. Now, in order to make it easy to carry out such requests, let us create a template for the client.

Client boilerplate

Let us now write the client boilerplate. This will be very similar to writing the server boilerplate but instead of creating a listener on an IP and port, we are going to call the `grpc.Dial` function and pass the connection options to it.

Once again, we are not going to hardcode the address we are going to connect to. We are going to take that as a parameter:

```
args := os.Args[1:]

if len(args) == 0 {
  log.Fatalln("usage: client [IP_ADDR]")
}

addr := args[0]
```

After that, we are going to create an instance of `DialOption`, and to keep this boilerplate generic, we are going to make an insecure connection to the server with the `insecure.NewCredentials()` function. Do not worry though; we will discuss how to make secure connections later:

```
opts := []grpc.DialOption{
  grpc.WithTransportCredentials(insecure.NewCredentials()),
}
```

Finally, we can just call for the `grpc.Dial` function to create a `grpc.ClientConn` object. This is the object that we are going to need later to call the API endpoints. Lastly, this is a connection object, so at the end of our client's lifetime, we are going to close it:

```
conn, err := grpc.Dial(addr, opts...)

if err != nil {
  log.Fatalf("did not connect: %v", err)
}

defer func(conn *grpc.ClientConn) {
  if err := conn.Close(); err != nil {
    log.Fatalf("unexpected error: %v", err)
  }
}(conn)
```

That is pretty much it for the client. The complete code is the following (`client/main.go`):

```
package main
```

```go
import (
  "log"
  "os"

  "google.golang.org/grpc"
  "google.golang.org/grpc/credentials/insecure"
)

func main() {
  args := os.Args[1:]

  if len(args) == 0 {
    log.Fatalln("usage: client [IP_ADDR]")
  }

  addr := args[0]
  opts := []grpc.DialOption{
    grpc.WithTransportCredentials(insecure.NewCredentials()),
  }
  conn, err := grpc.Dial(addr, opts...)

  if err != nil {
    log.Fatalf("did not connect: %v", err)
  }

  defer func(conn *grpc.ClientConn) {
    if err := conn.Close(); err != nil {
      log.Fatalf("unexpected error: %v", err)
    }
  }(conn)
}
```

Clearly, right now, it is not doing anything; however, we can test it by running our server first:

```
$ go run server/main.go 0.0.0.0:50051
```

Then, we run our client:

```
$ go run client/main.go 0.0.0.0:50051
```

The server should wait infinitely and the client should be returning without any error on the terminal. If this is the case, you are ready to write some gRPC API endpoints.

Bazel

This time, Bazel's setup will not be as long as for the server. This is mostly because we already have the deps.bzl file and we can reuse it for the client. All we need to do is use Gazelle to generate our BUILD.bazel and we are done:

```
$ bazel run //:gazelle
```

We should now have a BUILD.bazel in the client directory. The most important thing to note is that in this file, we can see Bazel linked gRPC to the client_lib. We should have something like this:

```
go_library(
  name = "client_lib",
  srcs = ["main.go"],
  deps = [
    "@org_golang_google_grpc//:go_default_library",
    "@org_golang_google_grpc//credentials/insecure",
  ],
  #...
)
```

We can now run our client in an equivalent way as the go run command:

```
$ bazel run //client:client 0.0.0.0:50051
```

We now have both our server and client. As of now, they do not do anything, but this is the intended purpose. Later in this book, by just copying them, we will be able to focus only on what matters, which is the API. Before doing any of that though, let us have a quick look at some of the most important options for the server and client setup.

Server and Dial options

We touched upon ServerOption and DialOption briefly with the grpc.ServerOption object and the grpc.WithTransportCredentials function. However, there are a lot of other options you can choose from. For readability's sake, I will not go into detail about every one of them, but I want to present some major options that you will probably use. All ServerOptions can be found at the root of the grpc-go repository in the file called server.go (https://github.com/grpc/grpc-go/blob/master/server.go) and the DialOptions in the file called dialoptions.go (https://github.com/grpc/grpc-go/blob/master/dialoptions.go).

grpc.Creds

This is an option, on both the server and client sides, that we will use when we talk about securing APIs. For now, we saw that we can call `grpc.WithTransportCredentials` with an `insecure.NewCredentials` result, and this gives us an insecure connection. This means that none of the requests and responses are encrypted; anyone could intercept these messages and read them.

`grpc.Creds` lets us provide a `TransportCredentials` object instance, which is a common interface for all the supported transport security protocols, such as TLS and SSL. If we had a certificate file called `server.crt` and a key file called `server.pem`, we could create the following `ServerOption`:

```
certFile := "server.crt"
keyFile := "server.pem"
creds, err := credentials.NewServerTLSFromFile(certFile, keyFile)

if err != nil {
  log.Fatalf("failed loading certificates: %v\n", err)
}
opts = append(opts, grpc.Creds(creds))
```

Similarly, on the client side, we would have a **Certificate Authority** (**CA**) certificate and we would need to create the following `DialOptions` to be able to communicate with the server:

```
certFile := "ca.crt"
creds, err := credentials.NewClientTLSFromFile(
  certFile, "")

if err != nil {
  log.Fatalf("error while loading CA trust certificate:
    %v\n", err)
}
opts = append(opts, grpc.WithTransportCredentials(creds))
```

As of now, you do not worry too much about that. As I mentioned, we will use this later and see how to get the certificates.

grpc.*Interceptor

If you are not familiar with interceptors, these are pieces of code that are called before or after handling a request (server side) or sending a request (client side). The goal is generally to enrich the request with some extra information, but it can also be used to log requests or deny certain requests if they do not have the right headers, for example.

We will see later how to define interceptors, but imagine that we have a piece of code logging our requests and another one checking that the authorization header is set. We could chain these interceptors on the server side like so:

```
opts = append(opts, grpc.ChainUnaryInterceptor
  (LogInterceptor(), CheckHeaderInterceptor()))
```

Note that the order of these interceptors is important because they will be called in the order provided in the `grpc.ChainUnaryInterceptor` function.

For the client side, we could have the same kind of log interceptor and another one adding the authorization header with a cached value of the token needed for authentication with the server. This would give something like the following:

```
opts = append(opts, grpc.WithChainUnaryInterceptor
  (LogInterceptor(), AddHeaderInterceptor()))
```

Finally, note that you can use other functions to add these interceptors. Here are the other ones:

- `WithUnaryInterceptor` to set one unary RPC interceptor
- `WithStreamInterceptor` to set one stream RPC interceptor
- `WithChainStreamInterceptor` to chain multiple stream RPC interceptors

We saw two important options that are configurable on both the server and client sides. By using credentials, we can secure our communication between the communication actors, and by using interceptors, we can run arbitrary code before sending or receiving requests. Obviously, we just saw two options and there are many more on both sides. If you are interested in checking out all of them, I invite you to go to the GitHub repository linked at the beginning of this section.

Summary

In this chapter, we created templates for our future servers and clients. The goal was to write the boilerplate code and set up our build so that we can generate code and run our Go applications.

We saw that we can use protoc manually to generate Go code and use it with our application. We then saw that we can make the process a little bit smoother by using Buf to generate the code for us. Finally, we saw that we can use Bazel to both generate our code and run our application in a single step.

Finally, we saw that we can use multiple `ServerOptions` and `DialOptions` to tweak the server and client. We mostly looked at `grpc.Creds` and interceptors, but there are a lot more options that we can check in the `grpc-go` repository.

In the next chapter, we will see how to write each type of API provided in gRPC. We will start with unary APIs, then check server and client streaming APIs, and finally, see how to write bidirectional streaming endpoints.

Quiz

1. What is the advantage of using protoc manually?

    A. No setup needed; you only need to install protoc.

    B. Shorter generation commands.

    C. We can both generate Go code and run the application.

2. What is the advantage of using Buf?

    A. No setup needed; you only need to install protoc.

    B. Shorter generation commands.

    C. We can both generate Go code and run the application.

3. What is the advantage of using Bazel?

    A. No setup needed; you only need to install protoc.

    B. Shorter generation commands.

    C. We can both generate Go code and run the application.

4. What is an interceptor?

    A. An external piece of code that intercepts the payload of the communication

    B. A piece of code that runs in the server handler

    C. A piece of code that runs before or after handling or sending a request

Answers

1. A
2. B
3. C
4. C

5

Types of gRPC Endpoints

In this chapter, we are going to see the different types of gRPC endpoints that you can write. For each of the endpoints, we are going to understand what the idea behind the type of communication we are talking about is, and we are going to define an RPC endpoint in our Protobuf service. Finally, we are going to implement that endpoint and write a client to consume the endpoint. At the end of this chapter, the goal is to implement a TODO API that will let us create, update, delete, and list the tasks in our TODO list.

In this chapter, we're going to cover the following main topics:

- What the four types of RPC endpoints you can write are
- When to use each of them
- How to implement the endpoint's logic
- How to consume a gRPC endpoint

Technical requirements

You can find the code for this chapter in the folder called `chapter5` in the companion repo for this book at `https://github.com/PacktPublishing/gRPC-Go-for-Professionals/tree/main/chapter5`.

Using the template

> **Note**
> This section is only necessary if you want to work with the same architecture as the github repository. Continuing working on the code from the previous code is totally fine.

If you remember, the goal of the previous chapter was to create a template that we can use to create a new gRPC project. As we will start such a project now, we will need to copy the content of the `chapter4` folder into the `chapter5` folder. To do that, simply run the following commands:

```
$ mkdir chapter5
$ cp -R chapter4/* chapter5
```

We pass -R in our case because we want to recursively copy all the files in the chapter4 folder.

And finally, we can also clean the template a little bit. We can remove the dummy directory in the proto folder because this was just there for testing our code generation. Then, if you used bazel for chapter4, we can delete all the build folders that start with the bazel- prefix. To do that, we can just simply remove them as follows:

```
$ cd chapter5
$ rm -rf proto/dummy
$ rm -rf bazel-*
```

We are now ready to use this template and see the different API endpoints that we can write with gRPC.

A Unary API

> **Important note**
>
> In terms of the underlying protocol, as we mentioned in *Chapter 1*, *Networking Primer*, the Unary API uses Send Header followed by Send Message and Half-Close from the client side, and Send Message plus Send Trailer from the server side. If you need a refresher on these operations, I recommend you quickly check the *RPC operations* section in *Chapter 1* of this book. This will help get a sense of what is going on when you are calling this API endpoint.

The easiest and most familiar API endpoint that you can write is a unary endpoint. These roughly translate to GET, POST, and other HTTP verbs that you probably worked with in REST APIs. You send a request and you get a response. Generally, these endpoints will be the ones that you use the most often to represent the treatment of one resource. For example, if you write a login method, you just need to send LoginRequest and receive LoginResponse.

For this section, we are going to write an RPC endpoint called AddTask. As its name suggests, this endpoint will create a new task in the list. So before being able to do that, we need to define what a *task* is.

A task is an object that contains the following properties:

- id: A number that is an identifier of that task

- description: The actual task to be done and that the user reads

- done: Whether or not the task is already done

- due_date: When this task is due

And if we translate that to Protobuf code, we would have something like the following (the file in chapter5/proto/todo/v1 called todo.proto):

```
syntax = "proto3";

package todo.v1;

import "google/protobuf/timestamp.proto";

option go_package = "github.com/PacktPublishing/
  gRPC-Go-for-Professionals/proto/todo/v1";

message Task {
  uint64 id = 1;
  string description = 2;
  bool done = 3;
  google.protobuf.Timestamp due_date = 4;
}
```

Notice the use of a Timestamp type here. This is a well-known type that is provided with Protobuf under the google.protobuf package. We are using this type to represent a point in time in the future at which the task should be completed. We could have rewritten our own Date type or used the Date type defined in the googleapis repository (https://github.com/googleapis/googleapis/blob/master/google/type/date.proto) but Timestamp is sufficient for this API.

Now that we have our task, we can think about our RPC endpoint. We want our endpoint to receive a task, which should be inserted into the list, and to return the identifier for that task to the client:

```
message AddTaskRequest {
  string description = 1;
  google.protobuf.Timestamp due_date = 2;
}

message AddTaskResponse {
  uint64 id = 1;
}

service TodoService {
  rpc AddTask(AddTaskRequest) returns (AddTaskResponse);
}
```

One important thing to notice here is that instead of using the Task message as the parameter for the AddTask endpoint, we created a wrapper (AddTaskRequest) around the necessary pieces of information for the endpoint. Nothing more, nothing less. We could have used the message directly but this may have led the client to send unnecessary data over the wire (e.g., setting the ID, which would be overlooked by the server). Furthermore, for future versions of our API, we could add more fields to `AddTaskRequest` without affecting the `Task` message. We effectively decouple the actual data representation for the request/response representation.

Code generation

> **Important note**
>
> If you are unsure by now on how to generate code out of a proto file, I highly recommend that you check *Chapter 4*, where we presented three methods for doing that. In this chapter, we are going to generate everything manually, but you can find how to do the same with Buf and Bazel in the `chapter5` folder of the GitHub repository.

Now that we have created the interface for our API endpoint, we want to be able to implement the logic behind it. The first step to do that is to generate some Go code. To do that, we are going to use `protoc` and the `source_relative` value for the `paths` option. So, knowing that our `todo.proto` file is under `proto/todo/v1/`, we can run the following command:

```
$ protoc --go_out=. \
         --go_opt=paths=source_relative \
         --go-grpc_out=. \
         --go-grpc_opt=paths=source_relative \
           proto/todo/v1/*.proto
```

After running that, you should have a `proto/todo/v1/` directory like the following:

```
proto/todo/v1/
├── todo.pb.go
├── todo.proto
└── todo_grpc.pb.go
```

That is all we need to get started.

Inspecting the generated code

In the generated code, we have two files – the Protobuf-generated code and the gRPC code. The Protobuf-generated code is in a file called `todo.pb.go`. If we inspect this file, the most important thing that we can see is the following code (which is simplified):

```
type Task struct {
    Id uint64
    Description string
```

```
  Done bool
  DueDate *timestamppb.Timestamp
}

type AddTaskRequest struct {
  Description string
  DueDate *timestamppb.Timestamp
}

type AddTaskResponse struct {
  Id uint64
}
```

This means we can now create `Task`, `TaskRequest`, and `TaskResponse` instances in our Go code, and that is exactly what we will do later in this chapter.

For the gRPC-generated code (`todo_grpc.pb.go`), there are interfaces generated for both the client and the server. They should look like the following:

```
type TodoServiceClient interface {
  AddTask(ctx context.Context, in *AddTaskRequest, opts
    ...grpc.CallOption) (*AddTaskResponse, error)
}

type TodoServiceServer interface {
  AddTask(context.Context, *AddTaskRequest)
    (*AddTaskResponse, error)
  mustEmbedUnimplementedTodoServiceServer()
}
```

They look similar, but the server-side `AddTask` is the only one that we need to implement the logic for. The client-side `AddTask` basically generates the request for calling the API endpoint on our server and returning us the response that it received.

Registering a service

To let gRPC know how to handle a certain request, we need to register the implementation of a service. To register such a service, we are going to call a generated function, which is present in `todo_grpc.pb.go`. In our case, this function is called `RegisterTodoServiceServer` and its function signature is the following:

```
func RegisterTodoServiceServer(s grpc.ServiceRegistrar, srv
  TodoServiceServer)
```

It takes `grpc.ServiceRegistrar`, which is an interface that `grpc.Server` implements, and takes `TodoServiceServer`, which is the interface that we saw earlier. This function will link the generic gRPC server provided by the framework with our implementation of the endpoints so that the framework knows how to handle requests.

So, the first thing to do is to create our server. We are first going to create a struct that embeds `UnimplementedTodoServiceServer`, which is a generated struct that contains the default implementation of the endpoints. In our case, the default implementation is the following:

```
func (UnimplementedTodoServiceServer) AddTask
  (context.Context, *AddTaskRequest) (*AddTaskResponse,
    error) {
  return nil, status.Errorf(codes.Unimplemented, "method
    AddTask not implemented")
}
```

If we do not implement `AddTask` in our server, this endpoint will be called, and it will return an error every time we call it. Right now, that does not seem useful because this does nothing other than return an error, but the fact is that this is a safety net, and we are going to see why when we talk about evolving our APIs.

Next, our server will contain a reference to our database. You can adapt this to use any database you are familiar with, but in our case, we are going to abstract the database away with an interface because this will let us focus on gRPC and not on another technology such as, for example, MongoDB.

So, our server type (`server/server.go`) will look like this:

```
package main

import (
  pb "github.com/PacktPublishing/gRPC-Go-for-Professionals/proto/
    todo/v1"
)

type server struct {
  d db
  pb.UnimplementedTodoServiceServer
}
```

Now, let us see what the db interface looks like. We are going to first have a function called `addTask`, which takes a description and a `dueDate` value and returns the `id` value for the task created or an error. Now, the important thing to note is that this database interface should be decoupled from the generated code. This is, once again, due to the evolution of our API because if we were to change our endpoints or `Request`/`Response` objects, we would have to change our interface and all the implementations. Here, the interface is independent of the generated code. In `server/db.go`, we can now write the following:

```
package main

import "time"

type db interface {
  addTask(description string, dueDate time.Time) (uint64,
    error)
}
```

This interface will let us test against fake database implementations and implement an in-memory database for non-release environments.

The last step is to implement the in-memory database. We are going to have a regular array of `Task` storing our to-dos and `addTask` will simply append a task to that array and return the ID of the current task. In a file called `server/in_memory.go`, we can add the following:

```
package main

import (
  "time"
  pb "github.com/PacktPublishing/gRPC-Go-for-Professionals/
    proto/todo/v1"
  "google.golang.org/protobuf/types/known/timestamppb"
)

type inMemoryDb struct {
  tasks []*pb.Task
}

func New() db {
  return &inMemoryDb{}
}

func (d *inMemoryDb) addTask(description string, dueDate
  time.Time) (uint64, error) {
  nextId := uint64(len(d.tasks) + 1)
  task := &pb.Task{
    Id: nextId,
    Description: description,
    DueDate: timestamppb.New(dueDate),
  }

  d.tasks = append(d.tasks, task)
  return nextId, nil
}
```

There are a few things to note about this implementation. First, this is probably obvious, but this is not an optimal "database" and is only used for development purposes. Second, without going into too much detail, we could use Golang build tags to choose the database we wanted to run at compile time. For example, if we had our inMemoryDb and a mongoDb implementation, we could have go:build tags at the top of each file. For in_memory.go, we could have the following:

```
//go:build in_memory_db
//...
type inMemoryDb struct

func New() db
```

And for mongodb.go, we could have this:

```
//go:build mongodb
//...
type mongoDb struct

func New() db
```

This would let us select at compile time which New function we wanted to use and thus create either an inMemoryDb or mongoDb instance.

Finally, you might have noticed that we are using the generated code in the implementation of this "database." As this is a "database" that we use as a dev environment, it does not really matter if this implementation is coupled with our generated code. The most important thing is to not couple the db interface with it so that you can use any database without even having to deal with the generated code.

Now, we are finally ready to register our server type. To do that, we can just go into our server/main.go main function and add the following line:

```
import pb "github.com/PacktPublishing/gRPC-Go-for-Professionals/
    proto/todo/v1"
//...
s := grpc.NewServer(opts...)

pb.RegisterTodoServiceServer(s, &server{
  d: New(),
})

defer s.Stop()
```

This means that we have now linked the gRPC server called s to the server instance created. Note that the New function here is the function that we defined inside the in_memory.go file.

Implementing AddTask

For the implementation, we are going to create a file called `server/impl.go` that will contain the implementation of all our endpoints. Note that this is purely for convenience and that you could have a file per RPC endpoint.

Now, as you might remember, the generated interface for our server wants us to implement the following function:

```
AddTask(context.Context, *AddTaskRequest)
  (*AddTaskResponse, error)
```

So, we can just add that function to our server type by writing the function preface with the name of the server instance and the server type:

```
func (s *server) AddTask(_ context.Context, in
  *pb.AddTaskRequest) (*pb.AddTaskResponse, error) {
}
```

And finally, we can implement the function. This will call the `addTask` function from our db interface, and as this never returns errors (for now), we are going to take the given ID and return that as `AddTaskResponse`:

```
package main

import (
  "context"

  pb "github.com/PacktPublishing/gRPC-Go-for-Professionals/proto/
    todo/v1"
)

func (s *server) AddTask(_ context.Context, in
  *pb.AddTaskRequest) (*pb.AddTaskResponse, error) {
  id, _ := s.d.addTask(in.Description, in.DueDate.AsTime())

  return &pb.AddTaskResponse{Id: id}, nil
}
```

Note that `AsTime` is a function provided by the `google.golang.org/protobuf/types/known/timestamppb` package, which returns a Golang `time.Time` object. The `timestamppb` package is a collection of functions that lets us manipulate the `google.protobuf.Timestamp` object and use it in an idiomatic way in our Go code.

Right now, you might feel that this is too simplistic but remember that we are just at the beginning of our API. Later in the book, we will do error handling and see how to reject incorrect arguments.

Calling AddTask from a client

Finally, let us see how to call the endpoint from Go client code. This is simple since we already have the boilerplate that we created in the previous chapter.

We will create a function called `AddTask`, which will call the API endpoint that we registered in the server. To do so, we are going to need to pass an instance of `TodoServiceClient`, a description of the task, and a due date. We will create the client instance later but note that `TodoServiceClient` is the interface that we saw when we inspected the generated code. In `client/main.go`, we can add the following:

```
import  (
  //...
  google.golang.org/protobuf/types/known/timestamppb
  pb "github.com/PacktPublishing/gRPC-Go-for-Professionals/
  proto/todo/v1"
  //...
)


func addTask(c pb.TodoServiceClient, description string,
  dueDate time.Time) uint64 {
}
```

After that, with the parameters, we can just construct a new instance of `AddTaskRequest` and send it to the server.

```
func addTask(c pb.TodoServiceClient, description string,
  dueDate time.Time) uint64 {
  req := &pb.AddTaskRequest{
    Description: description,
    DueDate: timestamppb.New(dueDate),
  }
  res, err := c.AddTask(context.Background(), req)
  //...
}
```

Finally, we will receive either an `AddTaskResponse` or an error from our API call. If there is an error, we log that on the screen, and if there is not, we log and return the ID:

```
func addTask(c pb.TodoServiceClient, description string,
  dueDate time.Time) uint64 {
  //...
  if err != nil {
    panic(err)
  }
```

```
fmt.Printf("added task: %d\n", res.Id)
return res.Id
}
```

To call this function, we need to use a generated function called `NewTodoServiceClient`, to which we pass a connection, and it returns a new instance of `TodoServiceClient`. And with that, we can simply add the following lines to the end of our `main` function in the client:

```
conn, err := grpc.Dial(addr, opts...)

if err != nil {
  log.Fatalf("did not connect: %v", err)
}
c := pb.NewTodoServiceClient(conn)

fmt.Println("--------ADD--------")
dueDate := time.Now().Add(5 * time.Second)
addTask(c, "This is a task", dueDate)
fmt.Println("------------------")

defer func(conn *grpc.ClientConn) {
  /*...*/}(conn)
```

Note that here we are adding a task with a five-second due date and a description of `This is a task`. This is just an example and I encourage you to try to make more calls by yourself with different values.

Now, we can basically run our server and client and see how they are interacting. To run the server, use this `go run` command:

```
$ go run ./server 0.0.0.0:50051
listening at 0.0.0.0:50051
```

And then, on another terminal, we can run the client in a similar way:

```
$ go run ./client 0.0.0.0:50051
--------ADD--------
added task: 1
------------------
```

And finally, to kill the server, you can just press *Ctrl* + *C* in the terminal running it.

So, we can see that we have a service implementation registered to our server, and our client is correctly sending a request and getting a response back. What we made is a simple example of a Unary API endpoint.

Bazel

> **Important note**
>
> In this section, we will see how to run an application with Bazel. However, because it would become repetitive if every section had such an explanation, I wanted to warn you that we will not go through these steps each time. For each section, you can run the `bazel run` commands (for the server and client) that you will see under, and the `gazelle` command should be useful only for this section.

At this point, your Bazel BUILD files are probably out of date. To synchronize them, we can simply run the `gazelle` command and it will update all our dependencies and files to compile:

```
$ bazel run //:gazelle
```

After that, we should be able to run the server and client easily by running this command to execute the server:

```
$ bazel run //server:server 0.0.0.0:50051
listening at 0.0.0.0:50051
```

Meanwhile, use the following to run the client:

```
$ bazel run //client:client 0.0.0.0:50051
--------ADD--------
added task: 1
-------------------
```

This is another example of how our server and client are working properly. We can run them with both the `go run` and `bazel run` commands. We are now confident with the Unary API; let us move to the server streaming API.

The server streaming API

> **Important note**
>
> In terms of the underlying protocol, the server streaming API uses `Send Header` followed by `Send Message` and `Half-Close` from the client side, and multiple `Send Message` plus `Send Trailer` from the server side.

Now, that we know how to register a service, interact with a "database," and run our client and server, everything will be faster. We will focus mostly on the API endpoint itself. In our case, we are going to create a `ListTasks` endpoint, which, as its name suggests, lists all the available tasks in the database.

One thing that we are going to do to make this a little bit fancier is that for each task listed, we are going to return whether this task is overdue or not. This is mostly done so that you can see how to provide more information about a certain object in the `response` object.

So, in the `todo.proto` file, we are going to add an RPC endpoint called `ListTasks`, which will take `ListTasksRequest` and return a stream of `ListTasksResponse`. This is what a server streaming API is. We get one request and return zero or more responses:

```
message ListTasksRequest {
}

message ListTasksResponse {
  Task task = 1;
  bool overdue = 2;
}

service TodoService {
  //...
  rpc ListTasks(ListTasksRequest) returns (stream ListTasksResponse);
}
```

Note that this time, we send an empty object as a request, and we get multiple `Task` and whether they are overdue. We could make the request a bit more intelligent by sending a range of IDs of tasks we want to list (paging), but for the sake of conciseness, we have chosen to make it simple.

Evolving the database

Before being able to implement the `ListTasks` endpoint, we need a way to access all the elements in our TODO list. Once again, we do not want to tie the `db` interface to our generated code, so we have a few choices:

- We create some abstractions to iterate over the tasks. This might be fine for our in-memory database, but how would this work with Postgres, for example?

- We tie our interface to an already existing abstraction such as cursors for databases. This is a bit better, but we still couple our interface.

- We simply leave the iteration to our implementation of the `db` interface and we apply a user-provided function to all the rows. With this, we are not coupled to any other component.

So, we are going to leave iteration to the implementation of our interface. This means that the new function that we will add to `inMemoryDb` will iterate over all the tasks and apply a function given as a parameter to each:

```
type db interface {
  //...
```

```
    getTasks(f func(interface{}) error) error
}
```

As you can see, the function passed as a parameter will itself get an `interface{}` as a parameter. This is not type-safe; however, we are going to make sure that we receive a `Task` at runtime when dealing with it later.

Then, for the in-memory implementation, we have the following:

```
func (d *inMemoryDb) getTasks(f func(interface{}) error)
  error {
  for _, task := range d.tasks {
    if err := f(task); err != nil {
      return err
    }
  }
  return nil
}
```

There is one thing to notice here. We are going to make errors fatal to the process. If the user-provided function returns an error, the `getTasks` function will return an error.

That is it for the database; we can now get all the tasks from it and apply some kind of logic to the tasks. Let us implement `ListTasks`.

Implementing ListTasks

To implement the endpoint, let us generate the Go code out of the proto files and take the function signature that we need to implement. We just run the following:

```
$ protoc --go_out=. \
        --go_opt=paths=source_relative \
        --go-grpc_out=. \
        --go-grpc_opt=paths=source_relative \
        proto/todo/v1/*.proto
```

And if we look at `proto/todo/v1/todo_grpc.pb.go`, we can now see that the `TodoServiceServer` interface has one more function:

```
type TodoServiceServer interface {
  //...
  ListTasks(*ListTasksRequest, TodoService_ListTasksServer)
    error
  //...
}
```

As we can see, the signature changed a bit from the one we had from the AddTask function. We now only return an error or `nil`, and we have `TodoService_ListTasksServer` as a parameter.

If you dig deeper into the generated code, you will see that `TodoService_ListTasksServer` is defined as follows:

```
type TodoService_ListTasksServer interface {
  Send(*ListTasksResponse) error
  grpc.ServerStream
}
```

This is a stream on which you can send `ListTasksResponse` objects.

Now that we know that, let us implement the function in our code. We can go to the `impl.go` file under the server and copy and paste the function signature of `ListTasks` into `TodoServiceServer`:

```
func (s *server) ListTasks(req *pb.ListTasksRequest, stream
  pb.TodoService_ListTasksServer) error
```

Obviously, we added the server type to specify that we are implementing `ListTasks` for our server and we named the parameters. `req` is the request that we receive from the client and `stream` is the object that we are going to use to send multiple answers.

Then, the logic of our function is once again straightforward. We are going to loop over all tasks, making sure that we are dealing with `Task` objects, and for each of these tasks, we are going to "calculate" the overdue by checking whether these tasks are done (with no overdue for tasks that are done) and whether the due_date field is before the current time. In summary, we will just create `ListTasksResponse` with this information and send that to the client (`server/impl.go`):

```
func (s *server) ListTasks(req *pb.ListTasksRequest, stream
  pb.TodoService_ListTasksServer) error {
  return s.d.getTasks(func(t interface{}) error {
    task := t.(*pb.Task)
    overdue := task.DueDate != nil && !task.Done &&
      task.DueDate.AsTime().Before(time.Now().UTC())
    err := stream.Send(&pb.ListTasksResponse{
      Task: task,
      Overdue: overdue,
    })
    return err
  })
}
```

One thing to notice is that the `AsTime` function will create a time in the UTC time zone, so when you compare a time with it you need it to also be in UTC. That is why we have `time.Now().UTC()` and not simply `time.Now()`

There are obviously ways in which this function can fail (e.g., what if the variable called t is not a Task?) but right now, let us not worry too much about error handling. We are going to see that later. Let us now call this endpoint from the client.

Calling ListTasks from a client

To call the ListTasks API endpoint from the client, we need to understand how to consume a server-streaming RPC endpoint. To do so, we check the method signature or the generated function in the interface name's TodoServiceClient. It should look like the following:

```
ListTasks(ctx context.Context, in *ListTasksRequest, opts
...grpc.CallOption) (TodoService_ListTasksClient, error)
```

We can see that we will need to pass a context and a request and that there are some optional call options. Then, we can also see that we will get a TodoService_ListTasksClient or an error. The TodoService_ListTasksClient type is very similar to the stream we dealt with in the ListTasks endpoint. The main difference though is that, instead of having a function called Send, we now have a function called Recv. Here is the definition of TodoService_ListTasksClient:

```
type TodoService_ListTasksClient interface {
    Recv() (*ListTasksResponse, error)
    grpc.ClientStream
}
```

So, what we are going to do with this stream is we are going to loop over all the responses that we can get from Recv, and then at some point, the server will say: "I am done." This happens when we get an error that is equal to io.EOF.

We can create a function in client/main.go called printTasks that will repeatedly call Recv and check whether we are finished or have an error, and if that is not the case, it will print the string representation of our Task object on the terminal:

```
func printTasks(c pb.TodoServiceClient) {
    req := &pb.ListTasksRequest{}
    stream, err := c.ListTasks(context.Background(), req)

    if err != nil {
        log.Fatalf("unexpected error: %v", err)
    }

    for {
        res, err := stream.Recv()

        if err == io.EOF {
```

```
    break
  }

  if err != nil {
    log.Fatalf("unexpected error: %v", err)
  }

  fmt.Println(res.Task.String(), "overdue: ",
    res.Overdue)
  }
}
```

Once we have this, we can call that function after the `addTask` call that we made in our `main` function:

```
fmt.Println("--------ADD--------")
//...

fmt.Println("--------LIST-------")
printTasks(c)
fmt.Println("------------------")
```

We can now use `go run` to run our server first and then our client. So, at the root of the project, we can run the following:

```
$ go run ./server 0.0.0.0:50051
listening at 0.0.0.0:50051
```

And then, we run our client:

```
$ go run ./client 0.0.0.0:50051
//...
--------LIST-------
id:1 description:"This is a task" due_date:
  {seconds:1680158076 nanos:574914000} overdue: false
------------------
```

This works as expected. Now, I would encourage you to try to add more tasks by yourself, try different values, and use the `printTasks` function after adding them all. This should help you get familiar with the API.

Now that we can add tasks and list them all, it would be nice if we could update already existing tasks. This might be interesting for marking a task as done and changing the due date. We are going to test that through the client streaming API.

The client streaming API

> **Important note**
>
> In terms of the underlying protocol, the client streaming API uses `Send Header` followed by multiple `Send Message` and a `Half-Close` from the client side, and `Send Message` plus `Send Trailer` from the server side.

With client streaming API endpoints, we can send zero or more requests and get one response. This is an important concept, especially for uploading data in real time. An example of this could be that we click on an edit button in our frontend, which triggers an edit session, and we post each edit being made in real time. Obviously, since we are not working with such fancy frontends, we are only going to focus on making the API compatible with this kind of feature.

To define a client streaming API, we simply need to write the `stream` keyword in the parameter clause instead of `return`. Previously, for our server streaming, we had the following:

```
rpc ListTasks(ListTasksRequest) returns (stream ListTasksResponse);
```

Now, we will have the following for `UpdateTasks`:

```
message UpdateTasksRequest {
  Task task = 1;
}

message UpdateTasksResponse {
}
service TodoService {
  //...
  rpc UpdateTasks(stream UpdateTasksRequest) returns
    (UpdateTasksResponse);
}
```

Note that, in this case, we are using the full Task message in the request, not separated fields like in `AddTask`. This is not a mistake and we will talk about this more during `chapter6`.

This effectively means that the client sends multiple requests, and the server returns one response. We are now one step closer to implementing our endpoint. However, let us talk about the database before doing that.

Evolving the database

Before thinking about implementing `UpdateTasks`, we need to see how we interact with the database. The first thing to consider is what information can be updated for a given task. In our case, we do not want the client to be able to update an ID; this is a detail that is handled by the database. However, for all the other information, we want to let the user be able to update it. When a task is done, we need to be able to set done to true. When the Task description needs updating, the client should be able to change it in the database. And finally, when the due date changes, the client should also be able to update it.

Knowing that, we can define our function signature for updateTask in our database. It will take the task ID and all the information that can be changed as parameters and return an error or nil:

```
type db interface {
  //...
  updateTask(id uint64, description string, dueDate
    time.Time, done bool) error
}
```

Once again, it looks a little bit much to pass that many parameters, but we do not want to couple this interface with any of the generated code. If later we need to add more information or remove some, this is as easy as updating this interface and updating the implementation.

Now, to implement that, we are going to go to the in_memory.go file. The function will simply iterate through all the tasks in the database, and if any Task has the same ID as the ID passed in the parameter, we will update all the fields one by one:

```
func (d *inMemoryDb) updateTask(id uint64, description
  string, dueDate time.Time, done bool) error {
  for i, task := range d.tasks {
    if task.Id == id {
      t := d.tasks[i]
      t.Description = description
      t.DueDate = timestamppb.New(dueDate)
      t.Done = done
      return nil
    }
  }

  return fmt.Errorf("task with id %d not found", id)
}
```

Now that means that every time we receive a request, we are going to iterate through all the tasks. This is not highly efficient, especially when the database becomes bigger. However, we are also not working with a real database so it should be good enough for our use case in this book.

Implementing UpdateTasks

To implement the endpoint, let us generate the Go code out of the proto files and take the function signature that we need to implement. We just run the following:

```
$ protoc --go_out=. \
         --go_opt=paths=source_relative \
         --go-grpc_out=. \
         --go-grpc_opt=paths=source_relative \
         proto/todo/v1/*.proto
```

And if we look at `proto/todo/v1/todo_grpc.pb.go`, we can now see that the `TodoServiceServer` interface has one more function:

```
type TodoServiceServer interface {
  //...
  UpdateTasks(TodoService_UpdateTasksServer) error
  //...
}
```

As we can see, the signature changed is similar to `ListTasks`; however, this time, we do not even deal with a request. We simply deal with a stream of the `TodoService_UpdateTasksServer` type. If we check the `TodoService_UpdateTasksServer` type definition, we have the following:

```
type TodoService_UpdateTasksServer interface {
  SendAndClose(*UpdateTasksResponse) error
  Recv() (*UpdateTasksRequest, error)
  grpc.ServerStream
}
```

We are already familiar with the `Recv` function. It lets us get an object, but now we also have a `SendAndClose` function. This function lets us tell the client that we are done on the server side. This is used to close the stream when the client sends an `io.EOF`.

Armed with that knowledge, we can implement our endpoint. We are going to repeatedly call the `Recv` function on the stream, and if we receive an `io.EOF`, we will use the `SendAndClose` function; otherwise, we will call the `updateTask` function on our database:

```
func (s *server) UpdateTasks(stream pb.TodoService
  _UpdateTasksServer) error {
  for {
    req, err := stream.Recv()

    if err == io.EOF {
      return stream.SendAndClose(&pb.UpdateTasksResponse{})
    }
```

```
    if err != nil {
      return err
    }

    s.d.updateTask(
      req.Task.Id,
      req.Task.Description,
      req.Task.DueDate.AsTime(),
      req.Task.Done,
    )
  }
}
```

We should now be able to trigger this API endpoint to change a given set of Task in real time. Let us now see how a client can call the endpoint.

Calling UpdateTasks from a client

This time, since we are working with client streaming, we are going to do the opposite from what we did for server streaming. The client will repeatedly call Send and the server will repeatedly call Recv. And in the end, the client will call the CloseAndRecv function, which is defined in the generated code.

If we look at the generated code for UpdateTasks on the client side, we will see the following signature in the TodoServiceClient type:

```
UpdateTasks(ctx context.Context, opts ...grpc.CallOption)
  (TodoService_UpdateTasksClient, error)
```

Notice that now the UpdateTasks function does not take any request parameters, but it will return a stream of the TodoService_UpdateTasksClient type. This type, as mentioned, will contain two functions: Send and CloseAndRecv. If we look at the generated code, we have the following:

```
type TodoService_UpdateTasksClient interface {
  Send(*UpdateTasksRequest) error
  CloseAndRecv() (*UpdateTasksResponse, error)
  grpc.ClientStream
}
```

Send is sent UpdateTasksRequest and CloseAndRecv will tell the server that it is done sending requests and ask for UpdateTasksResponse.

Now that we understand that we can implement the UpdateTasks function in the client/main. go file, we are going to call the UpdateTasks function from the gRPC client. This will return a stream and then we are going to send given tasks through it. Once we have looped through all the tasks that needed to be updated, we will call the `CloseAndRecv` function:

```go
func updateTasks(c pb.TodoServiceClient, reqs
...*pb.UpdateTasksRequest) {
  stream, err := c.UpdateTasks(context.Background())

  if err != nil {
    log.Fatalf("unexpected error: %v", err)
  }

  for _, req := range reqs {
    err := stream.Send(req)
    if err != nil {
      return
    }

    if err != nil {
      log.Fatalf("unexpected error: %v", err)
    }

    if req.Task != nil {
      fmt.Printf("updated task with id: %d\n", req.Task.Id)
    }
  }

  if _, err = stream.CloseAndRecv(); err != nil {
    log.Fatalf("unexpected error: %v", err)
  }
}
```

Now, as you can see in `updateTasks`, we need to pass zero or more `UpdateTasksRequest` as parameters. To get the IDs needed to create task instances and fill the `UpdateTasksRequest.task` field, we are going to record the IDs of tasks created before with `addTasks`. So, formerly, we had the following in `main`:

```go
addTask(c, "This is a task", dueDate)
```

We will now have something like the following:

```go
id1 := addTask(c, "This is a task", dueDate)
id2 := addTask(c, "This is another task", dueDate)
id3 := addTask(c, "And yet another task", dueDate)
```

And now, we can create an array of `UpdateTasksRequest` just like so:

```
[]*pb.UpdateTasksRequest{
  {Task: &pb.Task{Id: id1, Description: "A better name for
    the task"}},
  {Task: &pb.Task{Id: id2, DueDate: timestamppb.New
    (dueDate.Add(5 * time.Hour))}},
  {Task: &pb.Task{Id: id3, Done: true}},
}
```

This means that the `Task` object with `id1` will be updated to have a new description, the `Task` object with `id2` will be updated to have a new `due_date` value, and finally, the last one will be marked as `done`.

We can now pass the client to `updateTasks` and expand this array as a variadic parameter by using the ... operator. In `main`, we can now add the following:

```
fmt.Println("-------UPDATE------")
updateTasks(c, []*pb.UpdateTasksRequest{
  {Task: &pb.Task{Id: id1, Description: "A better name for
    the task"}},
  {Task: &pb.Task{Id: id2, DueDate: timestamppb.New
    (dueDate.Add(5 * time.Hour))}},
  {Task: &pb.Task{Id: id3, Done: true}},
}...)
printTasks(c)
fmt.Println("------------------")
```

We can now run that in a similar way to the previous sections. We use `go run` to run the server first:

```
$ go run ./server 0.0.0.0:50051
listening at 0.0.0.0:50051
```

And then we run the client to call the API endpoint:

```
$ go run ./client 0.0.0.0:50051
//...
-------UPDATE------
updated task with id: 1
updated task with id: 2
updated task with id: 3
id:1  description:"A better name for the task"  due_date:{}
id:2  due_date:{seconds:1680267768  nanos:127075000}
id:3  done:true  due_date:{}
------------------
```

Before moving on, there is an important thing to note here. You might be a little bit surprised about the fact that the tasks lost some information when updating them. This is because Protobuf will use the default value of a field when it is not set – meaning that if the client sends a `Task` object with only `done` being equal to `true`, the description will be deserialized as an empty string and `due_date` will be an empty `google.protobuf.Timestamp`. For now, this is highly inefficient because we need to resend all the information to update a single field. Later in the book, we are going to talk about how to solve this issue. We can now update multiple tasks in real time based on their IDs. Let us now move to the last API type available: bidirectional streaming.

The bidirectional streaming API

> **Important note**
>
> In terms of the underlying protocol, the bidirectional streaming API uses `Send Header` from the client side, followed by multiple `Send Message` from the server and/or client side, `Half-Close` from the client side, and finally `Send Trailer` from the server side.

In a bidirectional streaming API, the goal is to let the client send zero or more requests and let the server send zero or more responses. We are going to use this to simulate a feature that is like `updateTasks` but in which we are going to have direct feedback from the `DeleteTasks` endpoint API after each deletion instead of waiting for all the deletions to be done.

One thing to be clear about before continuing with the implementation is the question of why not to design `DeleteTasks` as a server streaming API or `updateTasks` as a bidirectional streaming API. The difference between these two tasks is how "destructive" they are. We can make an update directly on the client even before sending the requests to the server. If any error is present, we can simply look at the list we have on the client and synchronize it with the one on the server depending on the modified time. For a deletion, this would be a little bit more involved. We could keep the deleted row on the client and garbage would collect it later, or we would need to store the information somewhere else to synchronize it with the server. This creates a little bit more overhead.

So, we are going to send multiple `DeleteTasksRequest`, and for each, we are going to have the confirmation that it was deleted. And, if an error occurs, we are still sure that the tasks preceding the error were deleted on the server. Our RPC and messages look like the following:

```
message DeleteTasksRequest {
  uint64 id = 1;
}

message DeleteTasksResponse {
}

service TodoService {
```

```
//...
rpc DeleteTasks(stream DeleteTasksRequest) returns
  (stream DeleteTasksResponse);
}
```

We repeatedly send the ID of the Task we want to delete, and if we receive a DeleteTasksResponse, this means that the task was deleted. Otherwise, we get an error.

Now, before diving into the implementation, let us look at our database interface.

Evolving the database

As we want to remove a Task object in the database, we are going to need a deleteTask function. This function will take an ID of the Task object to be deleted, act on it, and return an error or nil. We can add the following function to server/db.go:

```
type db interface {
  //...
  deleteTask(id uint64) error
}
```

Implementing that looks a lot like updateTask. However, instead of updating information when we find the task with the right ID, we are going to delete it with a go slice trick. In server/in_memory.go, we now have the following:

```
func (d *inMemoryDb) deleteTask(id uint64) error {
  for i, task := range d.tasks {
    if task.Id == id {
      d.tasks = append(d.tasks[:i], d.tasks[i+1:]...)
      return nil
    }
  }

  return fmt.Errorf("task with id %d not found", id)
}
```

The slice trick takes the elements after the current task and appends them to the previous task. This effectively overrides the current task in the array and thus deletes it. With that, we are now ready to implement the DeleteTasks endpoint.

Implementing DeleteTasks

Before implementing the actual endpoint, we need to understand how we will deal with the generated code. So, let us generate the code with the following command:

```
$ protoc --go_out=. \
        --go_opt=paths=source_relative \
        --go-grpc_out=. \
        --go-grpc_opt=paths=source_relative \
        proto/todo/v1/*.proto
```

And if we check `todo_grpc.pb.go` in the `proto/todo/v1` folder, we should have the following function added to `TodoServiceServer`:

```
DeleteTasks(TodoService_DeleteTasksServer) error
```

This is similar to the `UpdateTasks` function because we get a stream, and we return either an error or nil. However, instead of having `Send` and `SendAndClose` functions, we now have `Send` and `Recv`. `TodoService_DeleteTasksServer` is defined like so:

```
type TodoService_DeleteTasksServer interface {
  Send(*DeleteTasksResponse) error
  Recv() (*DeleteTasksRequest, error)
  grpc.ServerStream
}
```

This means that, in our case, we can call `Recv` to get a `DeleteTasksRequest`, and for each of them we are going to send a `DeleteTasksResponse`. Finally, since we are working with a stream, we still need to check for errors and `io.EOF`. When we get `io.EOF`, we will just end the function with `return`:

```
func (s *server) DeleteTasks(stream
  pb.TodoService_DeleteTasksServer) error {
  for {
    req, err := stream.Recv()

    if err == io.EOF {
      return nil
    }

    if err != nil {
      return err
    }

    s.d.deleteTask(req.Id)
    stream.Send(&pb.DeleteTasksResponse{})
  }
}
```

One thing to note here is the `stream.Send` call. Even though this is simple, this is what differentiates client streaming from bidirectional streaming. If we did not have that call, we would effectively send multiple requests from the client, and in the end, the server would return `nil` to close the stream. This would be exactly the same as `UpdateTasks`. But because of the `Send` call, we now have direct feedback after each deletion.

Calling UpdateTasks from the client

Now that we have our endpoint, we can call it from the client. Before we can do it though, we need to look at the generated code for `TodoServiceClient`. We should now have the following function:

```
DeleteTasks(ctx context.Context, opts ...grpc.CallOption)
  (TodoService_DeleteTasksClient, error)
```

Once again, this is similar to what we saw in the `ListTasks` and `UpdateTasks` functions because it returns a string that we can interact with. However, as you can guess, we can now use `Send` and `Recv`. `TodoService_DeleteTasksClient` looks like this:

```
type TodoService_DeleteTasksClient interface {
  Send(*DeleteTasksRequest) error
  Recv() (*DeleteTasksResponse, error)
  grpc.ClientStream
}
```

With that generated code and the underlying gRPC framework, we can now send multiple `DeleteTasksRequest` and get multiple `DeleteTasksResponse`.

Now, we are going to create a new function in `client/main.go` that will take variadic parameters of `DeleteTasksRequest`. Then, we are going to create a channel that will help us wait for the entire process of receiving and sending to finish. If we did not do that, we would return from the function before finishing. This channel will be used in a goroutine that will use `Recv` in the background. Once we receive an `io.EOF` in this goroutine, we are going to close the channel. Finally, we are going to go over all the requests and send them, and once we are done, we are going to wait for the channel to be closed.

This might seem a little bit abstract right now but think about the job that the client needs to do. It needs to use `Recv` and `Send` simultaneously; thus, we need some simple concurrent code:

```
func deleteTasks(c pb.TodoServiceClient, reqs
...*pb.DeleteTasksRequest) {
  stream, err := c.DeleteTasks(context.Background())

  if err != nil {
    log.Fatalf("unexpected error: %v", err)
  }
```

```go
waitc := make(chan struct{})

go func() {
  for {
    _, err := stream.Recv()

    if err == io.EOF {
      close(waitc)
      break
    }
    if err != nil {
      log.Fatalf("error while receiving: %v\n", err)
    }

    log.Println("deleted tasks")
  }
}()

for _, req := range reqs {
  if err := stream.Send(req); err != nil {
    return
  }
}
if err := stream.CloseSend(); err != nil {
  return
}

<-waitc
}
```

Finally, before running our server and client, let us call that function in the main function. We are going to delete all the tasks that we created with addTasks and prove that there are no more tasks by trying to print all of them:

```go
fmt.Println("-------DELETE------")
deleteTasks(c, []*pb.DeleteTasksRequest{
  {Id: id1},
  {Id: id2},
  {Id: id3},
}...)

printTasks(c)
fmt.Println("--------------------")
```

With that, we can run the server first:

```
$ go run ./server 0.0.0.0:50051
listening at 0.0.0.0:50051
```

And then we can run our client:

```
$ go run ./client 0.0.0.0:50051
//...
-------DELETE------
2023/03/31 18:54:21 deleted tasks
2023/03/31 18:54:21 deleted tasks
2023/03/31 18:54:21 deleted tasks
------------------
```

Notice here that, instead of having a single response like client streaming, we have three responses (three `deleted tasks`). This is because we get a response per request. We effectively implemented bidirectional streaming.

We implemented bidirectional streaming here, which let us get feedback for each request we sent to the server. With that, we can make sure that we update resources on the client side without having to wait for a response or error from the server. This is interesting for use cases like ours that need real-time updates.

Summary

In this chapter, we saw the different types of APIs we can write in gRPC. We saw that we can create similar API endpoints as the ones we are used to in REST APIs. These endpoints are called unary endpoints. Then, we saw that we can make server streaming APIs to let the server return multiple responses. Similarly, we saw that a client can return multiple requests with client streaming. And finally, we saw that we can "mix" server and client streaming to get bidirectional streaming.

Our current endpoints are simplistic and do not handle a lot of cases that are crucial for production-grade APIs.

In the next chapter, we will start seeing what we can improve at the API level. This will let us first focus on the usability of the API before diving deeper into all the aspects of production-grade APIs.

Quiz

1. What kind of API endpoint should you use when you want to send one request and receive one response?

    A. Bidirectional streaming

  B. Client streaming

  C. Unary

2. What kind of API endpoint should you use when you want to send zero or more requests and receive one response?

  A. Server streaming

  B. Bidirectional streaming

  C. Client streaming

3. What kind of API endpoint should you use when you want to send one request and receive zero or more responses?

  A. Server streaming

  B. Client streaming

  C. Bidirectional streaming

4. What kind of API endpoint should you use when you want to send zero or more requests and receive zero or more responses?

  A. Client streaming

  B. Bidirectional streaming

  C. Server streaming

Answers

1. C

2. C

3. A

4. B

Designing Effective APIs

While gRPC is performant, it is easy to make mistakes that will cost you in the long term or at scale. In this chapter, we are going to see the considerations that are important in order to design efficient APIs in gRPC. Since we are talking about API design, the considerations are going to be linked to Protobuf because, as you know by now, we define our types and endpoints in Protobuf.

In this chapter, we are going to cover the following topics:

- How to choose the right integer type

- Understanding the impact of field tags on the size of serialized data

- How to use field masks to solve the over-fetching problem

- Understanding how repeated fields can lead to a bigger payload than expected

Technical requirements

For this chapter, you will find the relevant code in the folder called `chapter6` in the accompanying GitHub repository (`https://github.com/PacktPublishing/gRPC-Go-for-Professionals/tree/main/chapter6`).

Choosing the right integer type

Protobuf is mostly performant because of its binary format and because of its representation of integers. While some types such as strings are serialized "as is" and prepended with the field tag, type, and length, numbers – especially integers – are generally serialized in way fewer bits than how they are laid out in your computer memory.

However, you might have noticed that I said "generally serialized." This is because if you chose the wrong integer type for your data, the `varint` encoding algorithm might encode an `int32` into 5 bytes or more, whereas, in memory, it is 4 bytes.

Let us see an example of a bad choice of integer type. Let us say that we want to encode the value 268,435,456. We can check how this value would be serialized in memory and with Protobuf by using the unsafe.Sizeof function from the Go standard library and the proto.Marshal function provided by Protobuf. And finally, we are also going to use the well-known Int32Value type to wrap the value and be able to serialize it with Protobuf.

Before writing the main function, let us try to make a generic function called serializedSize, which will return the size of an integer in memory and the size of the same integer being serialized with Protobuf.

> **Important note**
>
> The code presented here is present in the accompanying GitHub repository under the helpers directory. We thought it would not make sense to mix the TODO API and this kind of code so we separated it.

Let us first add the dependencies:

```
$ go get -u google.golang.org/protobuf
$ go get -u golang.org/x/exp/constraints
```

The first one is to have access to the well-known Int32Value type and the second one is to have access to predefined type constraints for generics.

We are going to use generics to accept any kind of integer as data and let us specify a wrapper message to be able to serialize the data with Protobuf. We will have the following function:

```
func serializedSize[D constraints.Integer, W
  protoreflect.ProtoMessage](data D, wrapper W) (uintptr,
    int) {
  //...
}
```

Then, we can simply use the proto.Marshal function from the Protobuf library to serialize the wrapper and return both the result of unsafe.Sizeof and the length of the serialized data:

```
func serializedSize[D constraints.Integer, W
  protoreflect.ProtoMessage](data D, wrapper W) (uintptr, int) {
  out, err := proto.Marshal(wrapper)

  if err != nil {
    log.Fatal(err)
  }

  return unsafe.Sizeof(data), len(out) - 1
}
```

After that, it is simple. We can just call that function from our `main` with a variable containing the value `268,435,456` and an instance of `Int32Value`:

```
import (
  "fmt"
  "unsafe"

  "google.golang.org/protobuf/proto"
  "google.golang.org/protobuf/reflect/protoreflect"
  "google.golang.org/protobuf/types/known/wrapperspb"

  "golang.org/x/exp/constraints"
)

//...

func main() {
  var data int32 = 268_435_456
  i32 := &wrapperspb.Int32Value{
    Value: data,
  }

  d, w := serializedSize(data, i32)

  fmt.Printf("in memory: %d\npb: %d\n", d, w)
}
```

If we run this, we should get the following result:

```
$ go run integers.go
in memory: 4
pb: 5
```

Now, if you looked carefully at the code, you might be thinking that the `-1` after `len(out)` is cheating. With Protobuf, `Int32Value` is serialized into 6 bytes. While you are right about the fact that the real serialization size is 6 bytes, the first bytes represent the type and field tag. So, to keep the comparison of the serialized data fair, we remove the metadata and only compare the number itself.

You may be thinking that our current TODO API, which uses `uint64` for IDs, also has this problem, and you would be totally right. You can easily see that by switching `int32` to `uint64`, `Int32Value` to `UInt64Value`, and setting our data to be equal to 72,057,594,037,927,936:

```
func main() {
  var data uint64 = 72_057_594_037_927_936
  ui64 := &wrapperspb.UInt64Value{
```

```
    Value: data,
  }

  d, w := serializedSize(data, ui64)

  fmt.Printf("in memory: %d\npb: %d\n", d, w)
}
```

With the preceding code, we would get the following result:

```
$ go run integers.go
in memory: 8
pb: 9
```

This means that after approximately 72 quadrillion tasks are registered, we will have this problem. Obviously, for our use case, we are safe using uint64 as id because to have such a problem we would need every person on the planet to create 9 million tasks (72 quadrillion / 8 billion). But this might problem might be more significant in other use cases, and we need to be aware of the limitations of our API.

An alternative to using integers

An alternative that is often cited and even recommended by Google is to use strings for IDs. They mention that 2^64 (int64) is not "as big as it used to be." In the context of the company, this is understandable. They must deal with a lot of data and with bigger numbers than a lot of us.

However, this is not the only advantage that a string has over a number type. The biggest advantage is probably the evolution of your API. If, at some point, you need to store bigger numbers, the only alternative you have is to switch to the string type. But the problem is that there is no backward and forward compatibility between the number type you used previously and a string. Thus, you will have to add a new field to your schema, clutter the message definition, and make the developers check whether the ID is set as a string or as a number in case of communication with older/newer versions of an application.

Strings also provide safety in the fact that these cannot be used for arithmetic operations. This limits smart developers, in a good way, to not being able to pull smart tricks with IDs and end up making the numbers overflow. IDs are effectively treated as globs that nobody should manually handle.

In conclusion, for some use cases, it might be a good idea to start directly with strings for IDs. If you expect to scale or simply deal with numbers that are bigger than the integer limits, a string is the solution. However, in a lot of cases, you will probably only need uint64. Just be aware of your needs and plan for the future.

Choosing the right field tag

As you know, field tags are serialized together with the actual data to let Protobuf know into which field to deserialize the data. And as these tags are encoded as `varint`, the bigger the tag, the bigger the impact on your serialized data size. In this section, let us talk about the two considerations that you must make to not let these tags affect your payload too much.

Required/optional

Having big field tags might be fine if you are aware of the trade-off. One common way of treating big tags is to see them as being used for optional fields. An optional field means that it is less often populated with data and because Protobuf does not serialize fields that are not populated, the tag itself is not serialized. However, we will occasionally populate this field and we will incur costs.

One advantage of such a design is keeping relevant information together without having to create loads of messages to keep the field tags small. It will make the code easier to read and make the reader aware of the possible fields that they can populate.

The downside though is that if you are creating an API that is user-facing, you might incur costs too often. This might be because the user does not understand how to use your API properly or simply because the user has specific needs. This might also happen in a company setting, but it can be mitigated by senior software engineers or internal documentation.

Let us see an example of the downside that big tags bring. For the sake of an example, let us say that we have the following message (`helpers/tags.proto`):

```
message Tags {
   int32 tag = 1;
   int32 tag2 = 16;
   int32 tag3 = 2048;
   int32 tag4 = 262_144;
   int32 tag5 = 33_554_432;
   int32 tag6 = 536_870_911;
}
```

Note that these numbers are not random. If you remember, during the Protobuf primer, I explained that tags are encoded as `varints`. These numbers are the thresholds for which it takes one more byte to serialize the tag alone.

Now, with that, we are going to calculate the size of the message into which we incrementally set the value for fields. We are going to start with an empty object, then we are going to set `tag`, then `tag2`, and so on. Note also that we are going to set the same value for all the fields (1). This will show us the overhead that it takes to simply serialize the tag.

In `helpers/tags.go`, we have the following:

```go
package main

import (
  "fmt"
  "log"

  "google.golang.org/protobuf/proto"
  "google.golang.org/protobuf/reflect/protoreflect"

  pb "github.com/PacktPublishing/gRPC-Go-for-Professionals/
    helpers/proto"
)

func serializedSize[M protoreflect.ProtoMessage](msg M) int
{
  out, err := proto.Marshal(msg)

  if err != nil {
    log.Fatal(err)
  }

  return len(out)
}

func main() {
  t := &pb.Tags{}
  tags := []int{1, 16, 2048, 262_144, 33_554_432, 536_870_911}
  fields := []*int32{&t.Tag, &t.Tag2, &t.Tag3, &t.Tag4,
    &t.Tag5, &t.Tag6}

  sz := serializedSize(t)
  fmt.Printf("0 - %d\n", sz)

  for i, f := range fields {
    *f = 1

    sz := serializedSize(t)
    fmt.Printf("%d - %d\n", tags[i], sz-(i+1))
  }
}
```

We repurposed the `serializedSize` we saw earlier. We set the field by dereferencing the pointer to the field, we calculate the size of the `Tag` message with the new field set, and we print the result. This result is a little bit manipulated to show us only the bytes for the tag. We subtract i+1 from the size because i is zero-indexed (so +1). So, effectively, we subtract the number of fields already set from the size, which is also the size it takes to serialize the data without the tag (1 byte for value 1).

In the end, if we run this, we have the following (beautified):

```
$ go run tags.go
Tag             Bytes
----            ----
0               0
1               1
16              3
2048            6
262144          10
33554432        15
536870911       20
```

This tells us that each time we pass a threshold, we get one more byte overhead in our serialized data. At first, we have an empty message, so we get 0 bytes, then we have a tag of 1, which is serialized into 1 byte, after that a tag of 2 serialized into 2 bytes, and so on. We can look at the difference between two lines to get the overhead. The overhead of setting `value` to a field with tag `2048` instead of setting it to a field with tag `16` is 3 bytes (6 – 3 bytes).

In conclusion, we need to keep the smaller field tags available for the fields that are the most populated or required. This is because these tags will almost always be serialized, and we want to minimize the impact of the tag serialization. For optional fields, we might use bigger tags to keep the related fields together, and with that, we should incur non-recurrent payload increases.

Splitting messages

In general, we prefer to split messages to keep smaller objects and have fewer fields, and thus smaller tags. This lets us arrange information into entities and understand what the given information is representing. Our `Task` message is an example of that. It groups information and we can reuse that entity in, for example, `UpdateTasksRequest` to accept a fully featured `Task` as a request.

However, while it is interesting to be able to separate information into entities, this does not come for free. Your payload gets affected by the use of a user-defined type. Let us see an example of splitting a message and how it can affect the size of serialized data. This example shows that there is a size overhead when splitting messages. To show that, we are going to create a message that contains a name and a wrapper around a name. This first time we check the size, we will only set the string, and the second time we will only set the wrapper. Here is what I mean by such a message:

```
message ComplexName {
```

```
      string name = 1;
   }

message Split {
   string name = 1;
   ComplexName complex_name = 2;
   }
```

Right now, let us not worry about the usefulness of this example. We are just trying to prove that splitting a message has an overhead.

Then, we will write a `main` function that simply sets the value to `name` first, then calculates the size and prints it. And then, we will clear the name, set the `ComplexName.name` field, calculate the size, and print it. If there is an overhead, the sizes should be different. In `helpers/split.go`, we have the following:

```
package main

import (
   "fmt"
   "log"

   "google.golang.org/protobuf/proto"
   "google.golang.org/protobuf/reflect/protoreflect"

   pb "github.com/PacktPublishing/gRPC-Go-for-Professionals
      /helpers/proto"
)

func serializedSize[M protoreflect.ProtoMessage](msg M) int
{
   out, err := proto.Marshal(msg)

   if err != nil {
      log.Fatal(err)
   }

   return len(out)
}

func main() {
   s := &pb.Split{Name: "Packt"}
   sz := serializedSize(s)
```

```
fmt.Printf("With Name: %d\n", sz)

s.Name = ""
s.ComplexName = &pb.ComplexName{Name: "Packt"}
sz = serializedSize(s)

fmt.Printf("With ComplexName: %d\n", sz)
}
```

If we run that, we should get:

```
$ go run split.go
With Name: 7
With ComplexName: 9
```

Effectively, these two sizes are different. But what is the difference? The difference is that user-defined types are serialized as length-delimited types. In our case, the simple name would be serialized as 0a 05 50 61 63 6b 74. 0a is the wire type for Length-Delimited + tag 1 and the rest are the characters. But for the complex type, we have 12 07 0a 05 50 61 63 6b 74. We recognize the last 7 bytes but there are two more in front. 12 is the `Length-Delimited wire type + tag 2` and 07 is the length of the following bytes.

In conclusion, we once again have a trade-off. The more tags we have in messages, the more possibility there is for us to incur costs in terms of payload size. However, the more we try to split messages to keep the tags small, the more we will also incur costs because the data will be serialized as length-delimited data.

Improving UpdateTasksRequest

To reflect on what we have learned in the last section, we are going to improve the serialized size of `UpdateTasksRequest`. This is important because of the context in which this message is used. This is a message that is sent 0 or more times by the client since it is used in a client streaming RPC endpoint. It means that any overhead in serialized data size will be multiplied by the number of times that we send this message over the wire.

> **Important note**
> The following code is present in the accompanying GitHub repository. You will find the new Protobuf code in the `proto/todo/v2` folder and the server/client code for `UpdateTasks` will be updated to reflect the change. Finally, one thing to notice is that we do not provide backward and forward compatibility. A server in `chapter6` cannot receive a request from a client in `chapter5`. More work is needed to make that possible.

If we look at the current message, we have the following:

```
message UpdateTasksRequest {
   Task task = 1;
}
```

This is describing exactly what we want, but now we know that some extra bytes will be serialized because of the sub-message. To solve this problem, we can simply copy the fields that we let the user change and the ID that describes which task to update. This will give us the following:

```
message UpdateTasksRequest {
   uint64 id = 1;
   string description = 2;
   bool done = 3;
   google.protobuf.Timestamp due_date = 4;
}
```

This is the same definition as the `Task` message.

Now, you might be thinking that we are repeating ourselves and that it is a waste to do so. However, there are two important benefits to doing that:

- We no longer need to incur overhead for the serialization of the user-defined type. On each request, we save 2 bytes (tag + type and length).

- We now have more control over the fields that a user might update. If we did not want the user to change `due_date` anymore, we would simply remove that from the `UpdateTaskRequest` message and reserve the tag 4.

To prove that this is more efficient in terms of serialized data size, we can temporarily modify the `UpdateTasks` function in `server/impl.go` a little bit for both `chapter5` and `chapter6`. To count the size of the payload, we can use the `proto.Marshal` that we used earlier and sum up the total serialized size. In the end, we can just print the result on the terminal when we receive an EOF.

Here is what it looks like in `chapter6`:

```
func (s *server) UpdateTasks(stream
   pb.TodoService_UpdateTasksServer) error {
   totalLength := 0

   for {
      req, err := stream.Recv()

      if err == io.EOF {
         log.Println("TOTAL: ", totalLength)
         return stream.SendAndClose(&pb.UpdateTasksResponse{})
```

```
    }

    if err != nil {
      return err
    }

    out, _ := proto.Marshal(req)

    totalLength += len(out)
    s.d.updateTask(
      req.Id,
      req.Description,
      req.DueDate.AsTime(),
      req.Done,
    )
  }
}
```

For `chapter5`, this leads to 56 bytes being sent over the network as requests, and for `chapter6`, we only send 50 bytes. Once again, this looks negligible because we are doing that at a small scale, but once we receive traffic, it will quickly pile up and impact our costs.

Adopting FieldMasks to reduce the payload

After improving our `UpdateTasksRequest` message, we can now start looking at `FieldMasks` to further reduce the payload size, but this time we are going to focus on `ListTasksResponse`.

First, let us understand what `FieldMasks` is. It refers to objects containing a list of paths telling Protobuf which fields to include and telling it implicitly which should not be included. An example of that could be the following. Saywe had a message such as `Task`:

```
message Task {
  uint64 id = 1;
  string description = 2;
  bool done = 3;
  google.protobuf.Timestamp due_date = 4;
}
```

And we wanted to select only `id` and `done` fields, we could have a simple `FieldMask` like the following:

```
mask {
  paths: "id"
  paths: "done"
}
```

We could then apply that mask on an instance of Task and it would keep only the mentioned fields' value. This is interesting when we are doing the equivalent of GET and we do not want to fetch too much unnecessary data (over-fetching).

Our TODO API contains one such use case: ListTasks. Why? Because if a user wanted to fetch only part of the information, they would not be able to do so. Selecting part of the data might be useful for features such as synchronizing tasks from local storage to a backend. If the backend has IDs 1, 2, and 3 and the local has 1, 2, 3, 4, and 5, we want to be able to calculate the delta of the tasks that we need to upload. To do this, we would need to list only the IDs as fetching the description, done date, and due_date value would be wasteful.

Improving ListTasksRequest

ListTasksResponse is a server-streaming kind of API. We send one request and we get 0 or more responses. This is important to mention because sending a FieldMask does not come for free. We still need to carry bytes on the wire. In our case, though, it is interesting to use masks because we can send it once and it will be applied to all the elements returned by the server.

The first thing that we need to do is to declare such a FieldMask. To do that, we import field_mask.proto and add a field to ListTasksRequest:

```
import "google/protobuf/field_mask.proto";
//...
message ListTasksRequest {
  google.protobuf.FieldMask mask = 1;
}
```

Then, we can go to the server side and apply that mask to all the responses that we send. This is done with reflection and a little bit of boilerplate. The first thing that we need to do is to add a dependency in the server to work with slices and specifically access the Contains function:

```
$ go get golang.org/x/exp/slices
```

After that, we can work with reflection. We are going to go over all the fields that a given message has and if its name is not present in the mask's paths, we are going to remove its value:

> **Important note**
>
> The following code is a simplistic implementation to filter fields in a message, but this is sufficient for our use case. In reality, there are more powerful features of FieldMasks such as filtering maps, lists, and sub-messages. Unfortunately, the Go implementation of Protobuf does not provide such utilities as the other implementations do, so we need to rely on writing our own code or using community projects.

```go
import (
  "google.golang.org/protobuf/proto"
  "google.golang.org/protobuf/reflect/protoreflect"
  "google.golang.org/protobuf/types/known/fieldmaskpb"
  "golang.org/x/exp/slices"
)

//...

func Filter(msg proto.Message, mask *fieldmaskpb.FieldMask) {
  if mask == nil || len(mask.Paths) == 0 {
    return
  }


  rft := msg.ProtoReflect()
  rft.Range(func(fd protoreflect.FieldDescriptor, _
    protoreflect.Value) bool {
    if !slices.Contains(mask.Paths, string(fd.Name())) {
      rft.Clear(fd)
    }
    return true
  })
}
```

With that, we can now basically use `Filter` in our `ListTasks` implementation to filter the `Task` object that will be sent in `ListTasksResponse`:

```go
func (s *server) ListTasks(req *pb.ListTasksRequest, stream
  pb.TodoService_ListTasksServer) error {
  return s.d.getTasks(func(t interface{}) error {
    task := t.(*pb.Task)

    Filter(task, req.Mask)

    overdue := task.DueDate != nil && !task.Done &&
      task.DueDate.AsTime().Before(time.Now().UTC())
    err := stream.Send(&pb.ListTasksResponse{
      Task: task,
      Overdue: overdue,
    })

    return err
  })
}
```

Notice that `Filter` is called before calculating `Overdue`. This is because if we do not include `due_date` in `FieldMask`, we assume that the user does not care about the overdue. In the end, the overdue will be false, not serialized, and thus not sent over the wire.

Then, we need to see how to use that on the client side. In this example, `printTasks` is going to print only IDs. We are going to receive `FieldMask` as a parameter of `printTasks` and add it to `ListTasksRequest`:

```
func printTasks(c pb.TodoServiceClient, fm *fieldmaskpb
  .FieldMask) {
  req := &pb.ListTasksRequest{
    Mask: fm,
  }
  //...
}
```

And finally, with `fieldmaskpb.New`, we first create a `FieldMask` with the path `id`. This function will check that `id` is a valid path in the message that we provide as the first argument. If there is no error, we can set the `Mask` field in our `ListTasksRequest` instance:

```
func main() {
  //...
  fm, err := fieldmaskpb.New(&pb.Task{}, "id")
  if err != nil {
    log.Fatalf("unexpected error: %v", err)
  }
  //...
  fmt.Println("--------LIST-------")
  printTasks(c, fm)
  fmt.Println("-------------------")
  //...
}
```

If we run that, we should have the following output:

```
--------LIST-------
id:1 overdue:false
id:2 overdue:false
id:3 overdue:false
-------------------
```

Note that `overdue` is still printed as `false`, but in our case, it can be overlooked because we print overdue in the `printTasks` function and the default value of overdue (bool) is false..

Beware the unpacked repeated field

The last consideration is not helpful for our TODO API but is worth mentioning. In Protobuf, we have different ways of encoding repeated fields. We have packed and unpacked repeated fields.

Packed repeated fields

To understand, let us see an example of a packed repeated field. Let us say that we have the following message:

```
message RepeatedUInt32Values {
  repeated uint32 values = 1;
}
```

It is a simple list of the uint32 scalar type. If we serialized this with the values 1, 2, and 3, we would get the following result:

```
$ cat repeated_scalar.txt | protoc --encode=
  RepeatedUInt32Values proto/repeated.proto | hexdump -C
0a 03 01 02 03
00000005
```

repeated_scalar.txt from the preceding command contains the following:

```
values: 1
values: 2
values: 3
```

This is an example of a packed repeated field because of how the field wraps multiple values. You might think that this is normal since this is a list, but we are going to see later that this is not always true.

To understand what "wraps multiple values" means, we need to take a closer look at the hexadecimal presented by hexdump. We have 5 bytes: 0a 03 01 02 03. As we know, a repeated field is serialized as a length-delimited type. So 0a is the combination of the type (varint) and field tag (1), 03 means that we have three elements in the list, and the rest are the actual values.

Unpacked repeated fields

However, serialized data for repeated fields is not always that compact. Let us look at an example of an unpacked repeated field. Let us say that we add the packed option with the value false for the field called values:

```
message RepeatedUInt32Values {
  repeated uint32 values = 1 [packed = false];
}
```

Now, if we run the same command with the same values, we should have the following result:

```
$ cat repeated_scalar.txt | protoc --encode=
RepeatedUInt32Values proto/repeated.proto | hexdump -C
08 01 08 02 08 03
00000006
```

We can see that we have a totally different way of serializing the data. This time, we repeatedly serialize `uint32`. Here, 08 stands for the type (`varint`) and tag (1), and you can see that it is present three times as we have three values. If we have more than two values in the repeated field, this is effectively adding a byte per value. In our case, we serialize the whole as 6 bytes instead of the 5 previously.

Now, you might be thinking that you will just not use the `packed` option and you should always have a `packed` field. You would be right for repeated fields acting on scalars but not on more complex types. For example, strings, bytes, and user-defined types will always be serialized as unpacked and there is no way to avoid that.

Let us take an example with a user-defined type. Say we have the following Protobuf code:

```
message UserDefined {
  uint32 value = 1;
}


message RepeatedUserDefinedValues {
  repeated UserDefined values = 1;
}
```

We can now try to run the following command:

```
$ cat repeated_ud.txt | protoc --encode=
RepeatedUserDefinedValues proto/repeated.proto | hexdump -C
0a 02 08 01 0a 02 08 02 0a 02 08 03
0000000c
```

`repeated_ud.txt` from the preceding command contains the following:

```
values: {value: 1}
values: {value: 2}
values: {value: 3}
```

We can see that we now have a combination of both the overhead that we had with sub-messages earlier in the chapter and on top of that our repeated field is unpacked. We have 0a and 02, which correspond to the sub-message itself, and the 08 + value, which corresponds to the field called `value`. As you can see, this is now wasting much more bytes.

Now, as this is impossible to avoid on complex types, it is incorrect to say that we should never use repeated fields on such types. This is a very useful concept, and it should be used with care, and we should be aware of its cost.

Summary

In this chapter, we saw the main considerations that we need to take into account when we design our APIs. Most of them were related to Protobuf since it is the interface of our API, and it handles serialization/deserialization. We saw that choosing the right integer type is important and can lead to problems in terms of payload size but also when we want to evolve our API.

After that, we saw that choosing the right field tag is also important. This is due to the fact that tags are serialized along with the data and that they are serialized as `varints`. So the bigger the tag, the bigger our payload.

Then, we saw how we can leverage `FieldMasks` to select the data that we need and avoid the over-fetching problem. While this is a concept that is not that developed in gRPC Go, other implementations use that extensively. This significantly reduces the payload that we send across the wire.

And finally, we saw that we need to be careful when using repeated fields in Protobuf. This is because if we use them on a complex type, we will waste some bytes. However, repeated fields should not be avoided because of that. Sometimes they are the right data structure. In the next chapter, we are going to cover how to make API calls efficient and secure.

Quiz

1. Why is it not always more optimal to use `varint` for integer types?

    A. No reason, they are always more optimal than fixed integers

    B. `varint` encoding serializes bigger numbers into a bigger amount of bytes

    C. `varint` encoding serializes smaller numbers into a bigger amount of bytes

2. How can we get the number of bytes a message will be serialized into?

    A. `proto.Marshal + len`

    B. `proto.UnMarshal + len`

    C. `len`

3. What kind of tag should we give to fields that are often populated?

    A. Bigger tags

    B. Smaller tags

4. What is the main problem of splitting messages to use smaller tags?

   A. We have overhead because sub-messages are serialized as length-delimited types

   B. No problem – this is the way to go

5. What is `FieldMask`?

   A. A collection of fields' paths telling us what data to exclude

   B. A collection of fields' paths telling us what data to include

6. When is a repeated field serialized as unpacked?

   A. When repeated fields are acting on scalar types

   B. Only when we use the packed option with the value `false`

   C. When repeated fields are acting on complex types

Answers

1. B
2. A
3. B
4. A
5. B
6. C

7

Out-of-the-Box Features

Since writing production-ready APIs is more complicated than sending requests and receiving responses, gRPC has a lot more to offer than the simple communication patterns we saw. In this chapter, we are going to see the most important features that we can use in order to make our APIs robust, efficient, and secure.

In this chapter, we are going to cover the following topics:

- Dealing with errors, cancellation, and deadlines

- Sending HTTP headers

- Encrypting data over the wire

- Providing extra logic with interceptors

- Balancing requests to different servers

By the end of the chapter, we will have learned about the most important features that come right out of the box when we use gRPC.

Technical requirements

For this chapter, you will find the relevant code in the folder called `chapter7` in the accompanying GitHub repository (`https://github.com/PacktPublishing/gRPC-Go-for-Professionals/tree/main/chapter7`).

In the last section, I will use Kubernetes to show client-side load balancing. I assume that you already have Docker installed and a Kubernetes cluster. This can be done any way you want, but I provide a Kind (`https://kind.sigs.k8s.io/`) configuration to spin up a cluster easily and locally. This configuration is situated under the `k8s` folder of `chapter7` and in the file called `kind.yaml`. Once Kind is installed, you can use it like so:

```
$ kind create cluster --config k8s/kind.yaml
```

And you can dispose of it by running the following command:

```
$ kind delete cluster
```

Handling errors

Up until now, we have not discussed potential errors that could appear within or outside of the business logic. This is obviously not great for a production-ready API, so we are going to see how to solve them. In this section, we are going to concentrate our efforts on the RPC endpoint called AddTask.

Before starting to code, we need to understand how errors work in gRPC, but this should not be hard because they are pretty similar to what we are used to in REST APIs.

Errors are returned with the help of a wrapper struct called Status. This struct can be built in multiple ways but the ones we are interested in this section are the following:

```
func Error(c codes.Code, msg string) error
func Errorf(c codes.Code, format string, a ...interface{}) error
```

They both take a message for the error and an error code. Let us focus on the codes since the messages are just strings describing the error. The status codes are predefined codes that are consistent across the different implementations of gRPC. It is similar to the HTTP codes such as 404 and 500, but the main difference is that they have more descriptive names and that they are much fewer codes than in HTTP (16 in total).

To see all these codes, you can head to the gRPC Go documentation (https://pkg.go.dev/google.golang.org/grpc/codes#Code). It contains good explanations for each of the errors and it is less ambiguous than HTTP codes, so do not be afraid. For this section though, we are interested in two common errors:

- InvalidArgument
- Internal

The first one indicates that the client has specified an argument that is not correct for the proper functioning of the endpoint. The second indicates that an expected property of the system is broken.

InvalidArgument is perfect for validating inputs. We are going to use that in AddTask to make sure that the description of a Task is not empty (a task without a description is useless) and that a due date is specified and not in the past. Note that we make the due date required, but if you wanted to make it optional, we could just check if the DueDate property in the request was nil and act accordingly:

```
import (
  //...
  "google.golang.org/grpc/codes"
  "google.golang.org/grpc/status"
```

```
}

func (s *server) AddTask(_ context.Context, in *pb.AddTaskRequest)
(*pb.AddTaskResponse, error) {
  if len(in.Description) == 0 {
    return nil, status.Error(
      codes.InvalidArgument,
      "expected a task description, got an empty string",
    )
  }

  if in.DueDate.AsTime().Before(time.Now().UTC()) {
    return nil, status.Error(
      codes.InvalidArgument,
      "expected a task due_date that is in the future",
    )
  }

  //...
}
```

These checks will make sure that we have only useful tasks in our database, and that our due dates are in the future.

Finally, we have another error that could come from the addTask function, which would be an error relayed from the database. We could do extensive checks to create more precise error codes depending on each database error, but in our case, for simplicity, we are simply going to say that any database error is an Internal error.

We are going to get the potential error from the addTask function and do something similar to what we did for our InvalidArgument, but this time, it will be an Internal code, and we are going to use the Errorf function to relay the details of the error:

```
func (s *server) AddTask(_ context.Context, in *pb.AddTaskRequest)
(*pb.AddTaskResponse, error) {
  //...
  id, err := s.d.addTask(in.Description,
  in.DueDate.AsTime())

  if err != nil {
    return nil, status.Errorf(
      codes.Internal,
      "unexpected error: %s",
      err.Error(),
```

```
    )
  }

  //...
}
```

Now, we are done with the server side. We can switch to the client side – if you didn't notice before, we are already "handling" errors in addTask. We have the following lines:

```
res, err := c.AddTask(context.Background(), req)
if err != nil {
  panic(err)
}
```

Of course, clients might do fancier error handling or even recovery, but our goal for now is to see that our server errors are correctly propagated to the client. To test the InvalidArgument error, we can simply try to add a Task without a description. At the end of main, we can add the following:

```
import (
  //...
  "google.golang.org/grpc/codes"
  "google.golang.org/grpc/status"
)

func main() {
  //...
  fmt.Println("-------ERROR-------")
  addTask(c, "", dueDate)
  fmt.Println("-------------------")
}
```

Then, we run our server:

```
$ go run ./server 0.0.0.0:50051
listening at 0.0.0.0:50051
```

And our client should return the expected error:

```
$ go run ./client 0.0.0.0:50051
-------ERROR-------
panic: rpc error: code = InvalidArgument desc = expected a task
description, got an empty string
```

Then, we can check the due date error by providing a Time instance that is in the past:

```
fmt.Println("-------ERROR-------")
```

```
// addTask(c, "", dueDate)
addTask(c, "not empty", time.Now().Add(-5 * time.Second))
fmt.Println("-------------------")
```

We should get the following:

```
$ go run ./client 0.0.0.0:50051
-------ERROR-------
panic: rpc error: code = InvalidArgument desc = expected a task due_
date that is in the future
```

And finally, we are not going to show the Internal error because this would make us create a fake error in the in-memory database, but understand that it will return the following:

```
$ go run ./client 0.0.0.0:50051
-------ERROR-------
panic: rpc error: code = Internal desc = unexpected error: <AN_ERROR_
MESSAGE>
```

Before finishing this section, it is important to also understand how we can check the type of an error and act accordingly. We are going to basically panic but with more readable messages. For example, imagine the situation in which we have the following code:

```
rpc error: code = InvalidArgument desc = expected a task due_date that
is in the future
```

Instead, we are going to print this:

```
InvalidArgument: expected a task due_date that is in the future
```

To do that, we are going to modify addTask such that if there is an error, we will try to convert it into a status with the FromError function – if the conversion is done correctly, we will print the error code and the error message, and if it did not convert into a status, we will just panic as before:

```
func addTask(c pb.TodoServiceClient, description string, dueDate time.
  Time) uint64 {
  //...
  res, err := c.AddTask(context.Background(), req)

  if err != nil {
    if s, ok := status.FromError(err); ok {
      switch s.Code() {
      case codes.InvalidArgument, codes.Internal:
        log.Fatalf("%s: %s", s.Code(), s.Message())
      default:
        log.Fatal(s)
```

```
      }
    } else {
      panic(err)
    }
  }

  //...
}
```

And now, after running the client with one of the errors we defined earlier, we can get the following:

```
$ go run ./client 0.0.0.0:50051
-------ERROR-------
InvalidArgument: expected a task due_date that is in the future
```

Bazel

> **Important note**
>
> The commands presented here are needed every time you update the imports related to gRPC. For the sake of simplicity, we show it only once in this chapter and we assume you will be able to do it in the other sections.

As we are going to add more dependencies throughout this chapter, we will need to update our BUILD files. If we try to run the server with Bazel, right now, we will get an error that says the following:

```
No dependencies were provided.
Check that imports in Go sources match importpath attributes in deps.
```

To solve this problem, we can just run the gazelle command, like so:

```
$ bazel run //:gazelle
```

And then, we will be able to run the server and client correctly:

```
$ bazel run //server:server 0.0.0.0:50051
listening at 0.0.0.0:50051

$ bazel run //client:client 0.0.0.0:50051
```

To conclude, we saw that we can create an error on the server side with the Error and Errorf functions from the status package. We have multiple error codes that we can choose from. We only saw two, but they are common ones. And finally, on the client side, we saw that we can act accordingly depending on the error code by transforming a Golang error into a Status and writing conditions based on the status code.

Canceling a call

When you want to stop a call depending on certain conditions or interrupt a long-lived stream, gRPC provides you with cancellation functions that you can execute at any time.

If you have worked with Go on any distributed system code or API before, you probably saw a type called `context`. This is the idiomatic way to provide request-scoped information and signal across the API's actors, and this is an important piece of gRPC.

If you did not pay attention, up until now, we used `context.Background()` every time we made a request. In the Golang documentation, this is described as returning "*a non-nil, empty Context. It is never cancelled, has no values, and has no deadline.*" As you can guess, this alone is not suitable for production-ready APIs for the following reasons:

- What if the user wants to kill the request early?
- What if the API call never returns?
- What if we need the server to be aware of global values (e.g., an authentication token)?

In this section, let us focus on the first question, and in the next two sections, we will answer the others.

To gain the ability to cancel a call, we will be using the `WithCancel` function from the `context` package (`https://pkg.go.dev/context#WithCancel`). This function will return the constructed context and a `cancel` function that we can execute to interrupt the call made with the context. So, now, instead of only using `context.Background()`, we will create a context like so:

```
ctx, cancel := context.WithCancel(context.Background())
defer cancel()
```

Notice that this is important to call the `cancel` function at the end of the function to release the resources associated with the context. To make sure that the function is called, we can use a `defer`. However, this does not mean we cannot call the function before the end of the function.

As an example, we will create a fictional requirement. It is fictional because we are going to improve/remove the code we will write in this section in the following ones. The fictional requirement is to cancel a `ListTasks` call once we get an overdue `Task`. We can agree that it does not make sense in terms of features but regardless, this section's goal is to try to cancel a call.

To implement such a feature, we will create the context with the `WithCancel` function, pass this context to the `ListTasks` API endpoint, and finally, we are going to add another `if` in the reading loop checking whether there is an `overdue`. If this is the case, we will call the `cancel` function:

```
func printTasks(c pb.TodoServiceClient) {
  ctx, cancel := context.WithCancel(context.Background())
  defer cancel()
```

```
//...
stream, err := c.ListTasks(ctx, req)
//...

for {
  //...

  if res.Overdue {
    log.Printf("CANCEL called")
    cancel()
  }

  fmt.Println(res.Task.String(), "overdue: ",
  res.Overdue)
  }
}
```

Note that we could be breaking instead of calling the `cancel` function directly. Then the `defer cancel()` would kick in and the server would stop working. However, I decided to directly call `cancel` and let the client-side loop run because I want to show you that we will receive an error when canceling a call.

Now, we need to be aware that `cancel` takes time to propagate over the network and thus the server might continue to run without us knowing. To inspect what the server is sending, we are simply going to print the `Task` on the terminal before sending it to the client:

```
func (s *server) ListTasks(req *pb.ListTasksRequest, stream
  pb.TodoService_ListTasksServer) error {
  return s.d.getTasks(func(t interface{}) error {
    //...

    log.Println(task)
    overdue := //...
    err := stream.Send(&pb.ListTasksResponse{
      //...
    })

    return err
  })
}
```

And finally, I want to mention that we do not need to add any code in the client's `main` and this is because we already saw that we have tasks that are overdue when we run the current `main` code. The first overdue should appear in the `update` section:

```
fmt.Println("-------UPDATE------")
updateTasks(c, []*pb.UpdateTasksRequest{
  {Id: id1, Description: "A better name for the task"},
  //...
}...)
printTasks(c, nil)
fmt.Println("------------------")
```

This is because, if you remember, we updated the Id value and the description of that Task and we set the rest of the properties of Task to the default values. This means that Done will be set to false and DueDate to an empty time object.

Now, we can run the server like so:

```
$ go run ./server 0.0.0.0:50051
listening at 0.0.0.0:50051
```

Then, we run the client:

> **Important note**
>
> Before running the following code, make sure that you have commented the function calls that panic. This includes the two addTask that we added in the previous section.

```
$ go run ./client 0.0.0.0:50051
```

You should notice, on the client side, that everything runs correctly and that nothing was canceled even if the update section contains the following message:

```
CANCEL called.
```

The reason for this is that, the server does not know that the call has been canceled. To solve this problem, we can make the server cancel-aware.

To do that, we need to check the context's Done channel. This channel will be closed when the cancel is propagated to the server and, in the cancel example, the context will have an error equal to context. Canceled. When we have this event, we know that the server needs to return and effectively stop working on the rest of the requests:

```
func (s *server) ListTasks(req *pb.ListTasksRequest, stream
  pb.TodoService_ListTasksServer) error {
  ctx := stream.Context()

  return s.d.getTasks(func(t interface{}) error {
    select {
```

```
    case <-ctx.Done():
      switch ctx.Err() {
      case context.Canceled:
        log.Printf("request canceled: %s", ctx.Err())
      default:
      }
      return ctx.Err()
    /// TODO: replace following case by 'default:' on production APIs.
    case <-time.After(1 * time.Millisecond):
    }


    //...
  })
}
```

There are a couple of things to note before running this code.

The first one is that, when dealing with `stream`, we can get the context by using the `Context()` function available in the generated stream type (in our case, `pb.TodoService_ListTasksServer`).

And secondly, note that we are intentionally sleeping 1 millisecond per call of the closure. This would not happen in production; we would have a default branch instead. This is done so that the server has time to notice the cancellation. Note that this number is arbitrary; it is the smallest amount of time for me to notice the cancel error on my machine. You might need to make it larger, or you can make it smaller.

Now, we can run the server like so:

```
$ go run ./server 0.0.0.0:50051
listening at 0.0.0.0:50051
```

Then, we can run the client:

```
$ go run ./client 0.0.0.0:50051
//...
CANCEL called
id:1 description:"A better name for the task" due_date:{}
overdue:  true
unexpected error: rpc error: code = Canceled desc = context canceled
```

Finally, you should also notice that on the server side, you get the following message:

```
request canceled: context canceled
```

To conclude, we saw how we can use `context.WithCancel()` to create a cancellable context. We also saw that this function returns a `cancel` function, which we need to call at the end of the scope to release the resources attached to the context, but we can also call it earlier to cancel depending on some condition. And finally, we saw that we can make the server cancel-aware so that it does not execute more work than needed.

Specifying deadlines

Deadlines are the most important thing when we are dealing with asynchronous communication. This is because a call could never return due to network or other problems. That is why Google recommends that we set a deadline for each of our RPC calls. Fortunately for us, this is as easy as canceling a call.

The first thing that we need to do, on the client side, is to create a context. This is similar to the `WithCancel` function, but this time, we will use `WithTimeout`. It takes a parent context like `WithCancel` but on top of that, it takes a `Time` instance representing the maximum amount of time for which we are willing to wait for a server answer.

Instead of `WithCancel` in `printTasks`, we are now going to have the following context:

```
ctx, cancel := context.WithTimeout(context.Background(), 1*time.
Millisecond)
defer cancel()
```

Obviously, a timeout of 1 millisecond is way too low to let the server answer, but this is done on purpose so that we can get a `DeadlineExceeded` error. In real-life scenarios, we would need to set the timeout depending on the requirements set for the service. This is very much dependent on your use case and the service's job, so you will need to experiment and track the average amount of time in which your server will respond.

That is everything we need to make a deadline in gRPC. We can now run our server:

```
$ go run ./server 0.0.0.0:50051
listening at 0.0.0.0:50051
```

Then, we can run the client:

```
$ go run ./client 0.0.0.0:50051
//...
unexpected error: rpc error: code = DeadlineExceeded desc = context
  deadline exceeded
```

We can see that the first `ListTasks`, as expected, failed.

Now, while this is all you need to set deadlines up, you can also make the server aware of the `DeadlineExceeded` error. Even if this is technically already done because we are returning `ctx.Err()` when the `Done` channel closes, we still want to print a message saying that the deadline has been exceeded.

To do so, this is like the `Canceled` error, but this time, we are going to add a `DeadlineExceeded` branch in the `switch` on `ctx.Err()`:

```go
func (s *server) ListTasks(req *pb.ListTasksRequest, stream
    pb.TodoService_ListTasksServer) error {
  ctx := stream.Context()

  return s.d.getTasks(func(t interface{}) error {
    select {
    case <-ctx.Done():
      switch ctx.Err() {
      //...
      case context.DeadlineExceeded:
        log.Printf("request deadline exceeded: %s",
        ctx.Err())
      }
      return ctx.Err()
    //...
  }

  //...
}
```

And if we rerun our server and client, we now should have the following message on the terminal running the server:

```
request deadline exceeded: context deadline exceeded
```

To conclude, we saw that, similarly to `WithCancel`, we can use `WithTimeout` in order to create a deadline for a call. It is recommended to always set a deadline since we might never get an answer back from the server. And finally, we also saw how to make the server deadline-aware so that it does not work more than needed.

Sending metadata

Another feature that builds upon contexts is the possibility to pass metadata through calls. In gRPC, these metadata can be HTTP headers or HTTP trailers. They are both a list of key-value pairs that are used for many purposes such as passing authentication tokens and digital signatures, data integrity, and so on. In this section, we are mostly going to focus on sending metadata through headers. Trailers are simply headers that are sent after a message and not before. They are less used by developers but are used by gRPC to implement streaming interfaces. Anyway, if you are interested, you can look at the `grpc.SetTrailer` function (`https://pkg.go.dev/google.golang.org/grpc#SetTrailer`).

Our use case will be to pass an auth token to the `UpdateTasks` RPC endpoint, and after checking it, we will decide to either update the task or return an `Unauthenticated` error. Obviously, we are not going to deal with how to generate the `auth` token because this is an implementation detail, but we are going to simply have `authd` as the right token and everything else will be considered incorrect.

Let us start from the server side. The server is receiving the data from the context; thus, we are going to use the `FromIncomingContext` function from the `metadata` package in gRPC. This will return a map and whether there is some metadata or not. In `UpdateTasks`, we can do the following:

```
func (s *server) UpdateTasks(stream pb.TodoService_UpdateTasksServer)
error {
  ctx := stream.Context()
  md, _ := metadata.FromIncomingContext(ctx)
  //...
}
```

We can now check whether the auth token was provided. To do that, this is a simple usage of the Golang map. We try to access the `auth_token` element in the `md` map. It will return values and a Boolean saying whether the key is in the map or not. If it is, we are going to check that there is only one value and that this value is equal to `"authd"`. And if it is not, we will return an `Unauthenticated` error:

```
func (s *server) UpdateTasks(stream pb.TodoService_UpdateTasksServer)
error {
  ctx := stream.Context()
  md, _ := metadata.FromIncomingContext(ctx)
  if t, ok := md["auth_token"]; ok {
    switch {
    case len(t) != 1:
      return status.Errorf(
        codes.InvalidArgument,
        "auth_token should contain only 1 value",
      )
    case t[0] != "authd":
      return status.Errorf(
        codes.Unauthenticated,
        "incorrect auth_token",
      )
    }
  } else {
    return status.Errorf(
      codes.Unauthenticated,
      "failed to get auth_token",
    )
  }
  //...
}
```

That is all for the server; we return an error if there is no metadata, if there is metadata but no auth_token, if there is an auth_token with multiple values, and if the auth_token value is different from "authd".

We can now go to the client to send the appropriate header. This can be done with another function from the metadata package called AppendToOutgoingContext. We know that we already have a context created before calling the endpoint, so we are just going to append auth_token to it. This is as easy as the following:

```go
func updateTasks(c pb.TodoServiceClient, reqs ...*pb.
  UpdateTasksRequest) {
  ctx := context.Background()
  ctx = metadata.AppendToOutgoingContext(ctx, "auth_token", "authd")
  stream, err := c.UpdateTasks(ctx)

  //...
}
```

We override the context that we created with a new context containing the key value pair. Note that the key-value pairs can be interleaved. This means that we can have the following:

```go
metadata.AppendToOutgoingContext(ctx, K1, V1, K2, V2, ...)
```

Here, K stands for key, and V stands for value.

We can now run the server:

```
$ go run ./server 0.0.0.0:50051
listening at 0.0.0.0:50051
```

And then we run the client:

```
$ go run ./client 0.0.0.0:50051
```

And everything should go well. However, if you set a value for auth_token different from "authd", you should get the following message:

```
unexpected error: rpc error: code = Unauthenticated desc = incorrect
auth_token
```

If you do not set the auth_token header, you will see this:

```
unexpected error: rpc error: code = Unauthenticated desc = failed to
get auth_token
```

And imagine you set multiple values to `auth_token`, like so:

```
ctx = metadata.AppendToOutgoingContext(ctx, "auth_token", "authd",
"auth_token", "authd")
```

You should get the following error:

```
unexpected error: rpc error: code = InvalidArgument desc = auth_token
should contain only 1 value
```

To conclude, we saw how to get metadata from a context with the `metadata.FromIncomingContext` function and all the possible errors that can arise when we do so. We also saw how to actually send metadata from the client by appending a key-value pair to a context with the `metadata.AppendToOutgoingContext` function.

External logic with interceptors

While some headers might be applicable to only one endpoint, most often, we want to be able to apply the same logic across different endpoints. In the case of the `auth_token` header, if we have multiple routes that can only be called when the user is logged in, we do not want to repeat all the checks we did in the previous section. It bloats the code; it is not maintainable; and it might distract developers when finding the heart of the endpoint. This is why we will use an authentication interceptor. We will extract that authentication logic and it will be called before each call in the API.

Our interceptors will be called `authInterceptor`. The interceptor on the server side will simply do all the checks we did in the previous section, and then if everything goes well, the execution of the endpoint will be launched. Otherwise, the interceptor will return the error and the endpoint will not be called.

To define a server side interceptor, we have two possibilities. The first one is used when we work with a unary RPC endpoint (e.g., `AddTasks`). The interceptor function will look like the following:

```
func unaryInterceptor(ctx context.Context, req interface{}, info
*grpc.UnaryServerInfo, handler grpc.UnaryHandler) (interface{}, error)
```

And then we have interceptors working on streams. They look like the following:

```
func streamInterceptor(srv interface{}, ss grpc.ServerStream, info
*grpc.StreamServerInfo, handler grpc.StreamHandler) error
```

They look very similar. The main difference is the parameters' type. Now, we are not going to use all the parameters for our use case, so I will encourage you to check the documentation (https://pkg.go.dev/google.golang.org/grpc) for `UnaryServerInterceptor` and `StreamServerInterceptor`, and play with them.

Let us start with the unary interceptor that will be used by `AddTasks`. We will first extract the check into a function that will be shared across the interceptors. In a file called `interceptors.go`, we can write the following:

```go
import (
  "context"

  "google.golang.org/grpc"
  "google.golang.org/grpc/codes"
  "google.golang.org/grpc/metadata"
  "google.golang.org/grpc/status"
)

const authTokenKey string = "auth_token"
const authTokenValue string = "authd"

func validateAuthToken(ctx context.Context) error {
  md, _ := metadata.FromIncomingContext(ctx)
  if t, ok := md[authTokenKey]; ok {
    switch {
    case len(t) != 1:
      return status.Errorf(
        codes.InvalidArgument,
        fmt.Sprintf("%s should contain only 1 value", authTokenKey),
      )
    case t[0] != authTokenValue:
      return status.Errorf(
        codes.Unauthenticated,
        fmt.Sprintf("incorrect %s", authTokenKey),
      )
    }
  } else {
    return status.Errorf(
      codes.Unauthenticated,
      fmt.Sprintf("failed to get %s", authTokenKey),
    )
  }

  return nil
}
```

There is nothing different from what we did directly in `UpdateTasks`. But now, writing our interceptor is simple. We will just call the `validateAuthToken` function and check for errors. If there is one, we will return it directly. And if there is not, we are going to call the `handler` function, which effectively calls the endpoint:

```go
func unaryAuthInterceptor(ctx context.Context, req interface{}, info
*grpc.UnaryServerInfo, handler grpc.UnaryHandler) (interface{}, error)
{
  if err := validateAuthToken(ctx); err != nil {
    return nil, err
  }

  return handler(ctx, req)
}
```

We can do the same for streams. The only thing that will change is the arguments of the handler and how we get the context:

```go
func streamAuthInterceptor(srv interface{}, ss grpc.ServerStream, info
*grpc.StreamServerInfo, handler grpc.StreamHandler) error {
  if err := validateAuthToken(ss.Context()); err != nil {
    return err
  }

  return handler(srv, ss)
}
```

Now, you might be thinking that we have functions but nobody to call them. And you are entirely right. We need to register these interceptors so that our server knows that they exist. This is done in `server/main.go` where we can add the interceptors as options to the gRPC server. Right now, we create the server like so:

```go
var opts []grpc.ServerOption
s := grpc.NewServer(opts...)
```

And to add the interceptors, we can simply add them to the `opts` variable:

```go
opts := []grpc.ServerOption{
  grpc.UnaryInterceptor(unaryAuthInterceptor),
  grpc.StreamInterceptor(streamAuthInterceptor),
}
```

We can now run the server:

> **Important note**
>
> Before running the server, you can delete the whole authentication logic in the `UpdateTasks` function in `server/impl.go`. This is not needed anymore since the interceptors will authenticate requests automatically.

```
$ go run ./server 0.0.0.0:50051
listening at 0.0.0.0:50051
```

Then, we can run the client:

```
$ go run ./client 0.0.0.0:50051
--------ADD--------
rpc error: code = Unauthenticated desc = failed to get auth_token
exit status 1
```

As expected, we get an error because we never added the `auth_token` header in the `addTask` function on the client side.

Obviously, we do not want to add the header to all our calls by hand. We will create a client-side interceptor that will add it for us before proceeding to send the request. On the client side, we have two ways of defining an interceptor. For the unary calls, we have the following:

```
func unaryInterceptor(ctx context.Context, method string, req
interface{}, reply interface{}, cc *grpc.ClientConn, invoker grpc.
UnaryInvoker, opts ...grpc.CallOption) error
```

And for streams, we have this:

```
func streamInterceptor(ctx context.Context, desc *grpc.StreamDesc, cc
*grpc.ClientConn, method string, streamer grpc.Streamer, opts ...grpc.
CallOption) (grpc.ClientStream, error)
```

As you can see, there are many more parameters on this side. And, while most parameters are not important for our use case, I encourage you to check the documentation (`https://pkg.go.dev/google.golang.org/grpc`) for `UnaryClientInterceptor` and `StreamClientInterceptor`, and play around with them.

In the client-side interceptors, we are going to simply create a new context and append the metadata before calling the endpoint. We do not even need to create a separate function to share the logic since this is as simple as calling the `AppendToOutgoingContext` function that we saw earlier.

In `client/interceptors.go`, we can write the following:

```
import (
```

```
    "context"

    "google.golang.org/grpc"
    "google.golang.org/grpc/metadata"
)

const authTokenKey string = "auth_token"
const authTokenValue string = "authd"

func unaryAuthInterceptor(ctx context.Context, method string, req,
    reply interface{}, cc *grpc.ClientConn, invoker grpc.UnaryInvoker,
        opts ...grpc.CallOption) error {
    ctx = metadata.AppendToOutgoingContext(ctx, authTokenKey,
        authTokenValue)
    err := invoker(ctx, method, req, reply, cc, opts...)

    return err
}

func streamAuthInterceptor(ctx context.Context, desc *grpc.StreamDesc,
    cc *grpc.ClientConn, method string, streamer grpc.Streamer, opts
        ...grpc.CallOption) (grpc.ClientStream, error) {
    ctx = metadata.AppendToOutgoingContext(ctx, authTokenKey,
        authTokenValue)
    s, err := streamer(ctx, desc, cc, method, opts...)

    if err != nil {
        return nil, err
    }

    return s, nil
}
```

And finally, as in the server, we also need to register these interceptors. This time, these interceptors will be registered by adding `DialOptions` to the `Dial` function we use in `main`. Right now, you should have something like:

```
opts := []grpc.DialOption{
    grpc.WithTransportCredentials(insecure.NewCredentials()),
}
```

We can now add the interceptors like so:

```
opts := []grpc.DialOption{
    //...
```

```
    grpc.WithUnaryInterceptor(unaryAuthInterceptor),
    grpc.WithStreamInterceptor(streamAuthInterceptor),
}
```

As they are registered, we can run our server:

```
$ go run ./server 0.0.0.0:50051
listening at 0.0.0.0:50051
```

Then, we can run the client:

> **Important note**
>
> Before running the client, you can delete the call to AppendToOutgoingContext present in updateTask in the client/main.go file. This is not needed anymore since the interceptors will do it automatically.

```
$ go run ./client 0.0.0.0:50051
```

And all the calls should now go through without any errors.

To conclude, in this section, we saw that we can write unary and stream interceptors on both the server and client sides. The goal of these interceptors is to automatically do some repetitive work across multiple endpoints. In our example, we automated the adding and checking of the auth_token header for authentication.

Compressing the payload

While Protobuf serializes data into binary and this involves much smaller payloads than text data, we can apply compression on top of the binary. gRPC provides us with the gzip Compressor (https://pkg.go.dev/google.golang.org/grpc/encoding/gzip) and for more advanced use cases, lets us write our own Compressor (https://pkg.go.dev/google.golang.org/grpc/encoding).

Now, before diving into how to use the gzip Compressor, it is important to understand that lossless compression might result in a bigger payload size. If your payload does not contain repetitive data, which is what gzip detects and compresses, you will send more bytes than needed. So, you will need to experiment with a typical payload and see how gzip affects its size.

To show an example of that, I included in the helpers folder a file called gzip.go, which contains a helper function called compressedSize. This function returns the original size of the serialized data and its size after gzip compression:

```
func compressedSize[M protoreflect.ProtoMessage](msg M) (int, int) {
    var b bytes.Buffer
```

```
gz := gzip.NewWriter(&b)
out, err:= proto.Marshal(msg)

if err != nil {
  log.Fatal(err)
}

if _, err := gz.Write(out); err != nil {
  log.Fatal(err)
}

if err := gz.Close(); err != nil {
  log.Fatal(err)
}

return len(out), len(b.Bytes())
}
```

As this is a generic function, we can use it with any message. We can start with a message that would not be suitable to compress: Int32Value. So, in the main function of the file, we are going to create an instance of Int32Value, pass it through the compressedSize function, and we are going to print both the original and the new size:

```
func main() {
  var data int32 = 268_435_456
  i32 := &wrapperspb.Int32Value{
    Value: data,
  }

  o, c := compressedSize(i32)
  fmt.Printf("original: %d\ncompressed: %d\n", o, c)
}
```

And if we run this, we should get the following:

```
$ go run gzip.go
original: 6
compressed: 30
```

The compressed payload is five times bigger than the original one. That is definitely something to avoid in production. Now obviously, most of the time, we do not send such simple messages, so let us see a more concrete example. We are going to use the Task message that we defined earlier in the book:

```
syntax = "proto3";
```

```
package todo;

import "google/protobuf/timestamp.proto";

option go_package = "github.com/PacktPublishing/gRPC-Go-for-
Professionals/helpers/proto";

message Task {
  uint64 id = 1;
  string description = 2;
  bool done = 3;
  google.protobuf.Timestamp due_date = 4;
}
```

Then, we can compile it with the following command:

```
$ protoc --go_out=. \
        --go_opt=module=github.com/PacktPublishing/
              gRPC-Go-for-Professionals/helpers \
        proto/todo.proto
```

And after that, we can now create an instance of `Task` and pass it the `compressedSize` function to see the result of compression:

```
func main() {
  task := &pb.Task{
    Id: 1,
    Description: "This is a task",
    DueDate: timestamppb.New(time.Now().Add(5 * 24 *
    time.Hour)),
  }

  o, c := compressedSize(task)
  fmt.Printf("original: %d\ncompressed: %d\n", o, c)
}
```

And if we run it, we should get the following sizes:

```
$ go run gzip.go
original: 32
compressed: 57
```

This is better than the previous example, but this is still not efficient since we are sending more bytes than needed. So, in the cases we saw previously, it would not make sense to use gzip compression.

Lastly, let us see an example of when compression is useful. Let us say that most of our `Task` instances have long descriptions. For example, we could have something like so:

```
task := &pb.Task{
  //...
  Description: `This is a task that is quite long and requires a lot
  of work.
  We are not sure we can finish it even after 5 days.
  Some planning will be needed and a meeting is required.`,
  //...
}
```

Then, running the `compressedSize` function will give us the following sizes:

```
$ go run gzip.go
original: 192
compressed: 183
```

The lesson here is that we need to know our data before enabling gzip compression in gRPC. Now, let us see how to enable it.

On the server side (`server/main.go`), this is as easy as adding the following import:

```
_ "google.golang.org/grpc/encoding/gzip"
```

Notice that we add an underscore before it in order to avoid getting an error from the compiler saying that we are not using the import.

That is all for the server. On the client side, there is a little bit more code, but this is also simple. We can enable compression on all the RPC endpoints by adding `DialOption` or we can enable it for a single endpoint by adding `CallOption` (https://pkg.go.dev/google.golang.org/grpc#CallOption).

For the first option, we can simply add the following:

```
opts := []grpc.DialOption{
  //...
  grpc.WithDefaultCallOptions(grpc.UseCompres
  sor(gzip.Name))
}
```

gzip adds the same import as the one in the server without the preceding underscore.

And for adding compression per call, we can add `CallOption`. If we wanted to add gzip compression to `AddTask` calls, we would have the following:

```
res, err := c.AddTask(context.Background(), req, grpc.
UseCompressor(gzip.Name))
```

To conclude, we saw that always adding compression is not a good idea and we should only add it after testing it on our data. Then, we saw how we can register the gzip compressor on the server and client. And finally, we saw that we can enable compression globally or per call.

Securing connections

Up until now, we have not made our connections secure – we used insecure credentials. In gRPC, we can use TLS, mTLS, and ATLS connections. The first uses a one-way authentication where the client can verify the server's identity. The second one is a two-way communication where the server verifies the client's identity and the client verifies the server's. And finally, ATLS is similar to TLS but designed and optimized for Google's use.

mTLS and ATLS are worth exploring if you are working on smaller-scale communication or working with Google Cloud, respectively. If you are interested in mTLS, you should check the mTLS folder in the grpc-go GitHub repository: https://github.com/grpc/grpc-go/tree/master/examples/features/encryption/mTLS. And if you want to use ATLS, check out this link: https://grpc.io/docs/languages/go/alts/. However, in our case, we are going to see the most frequently used form of encryption, which is TLS.

To do so, we are going to need to create some self-signed certificates. Obviously, in production, these certificates will be automatically created with something such as Let's Encrypt. However, once these certificates are available, the overall settings are the same.

Now, for the sake of simplicity, we are going to download these certificates from the examples in the grpc-go repository. These certificates can also be found in the certs directory under the chapter7 folder. We first need to get the server certificate and its key:

```
$ curl https://raw.githubusercontent.com/grpc/grpc-go/master/examples/
data/x509/server_cert.pem --output server_cert.pem
$ curl https://raw.githubusercontent.com/grpc/grpc-go/master/examples/
data/x509/server_key.pem --output server_key.pem
```

And then we need to get the **Certificate Authority (CA)** certificate:

```
$ curl https://raw.githubusercontent.com/grpc/grpc-go/master/examples/
data/x509/ca_cert.pem --output ca_cert.pem
```

Now, we can start with the server. We will add the credentials as ServerOption because we want all our calls to be encrypted. To create the credentials, we can use a function called NewServerTLSFromFile from gRPC's credential package. It reads two files, the server certificate and the server key:

```
func main() {
    //...
    creds, err := credentials.NewServerTLSFrom
    File("./certs/server_cert.pem", "./certs/server_key.pem")
```

```
  if err != nil {
    log.Fatalf("failed to create credentials: %v", err)
  }
}
```

And once this is created, we can use the `grpc.Creds` function, which creates a `ServerOption`:

```
opts := []grpc.ServerOption{
  grpc.Creds(creds),
  //...
}
```

Let's see what happens when we now try to run the server:

```
$ go run ./server 0.0.0.0:50051
listening at 0.0.0.0:50051
```

Then, we run the client:

```
$ go run ./client 0.0.0.0:50051
--------ADD--------
rpc error: code = Unavailable desc = connection error: desc = "error
reading server preface: EOF"
```

We get an error that basically tells us that the client was not able to connect to the server. To solve this, we need to go to the client side and create the `DialOption` for the credentials.

This time, we will use the `NewClientTLSFromFile` function, which takes the CA certificate. For testing purposes, we will add the host URL as a second argument (the certificate domain is *.test. example.com).

```
creds, err := credentials.NewClientTLSFromFile("./certs/ca_cert.pem",
  "x.test.example.com")
if err != nil {
  log.Fatalf("failed to load credentials: %v", err)
}
```

And to add the credentials, we use a function called `WithTransportCredentials`, which creates a `DialOption`.

```
opts := []grpc.DialOption{
  grpc.WithTransportCredentials(creds),
  //grpc.WithTransportCredentials(insecure.NewCredentials())
  //...
}
```

Note that we remove the insecure credentials since we now want to encrypt the communication.

Let us now rerun the server:

```
$ go run ./server 0.0.0.0:50051
listening at 0.0.0.0:50051
```

Then, we do the same for the client:

```
$ go run ./client 0.0.0.0:50051
```

Everything goes well – we should pass through all the calls that we passed previously but now our communication is secure.

Bazel

In order to run the code we wrote in this section with Bazel, we need to include the certificate files in our BUILD files. This can be done by exporting them and adding them as data to the server_lib and client_lib targets.

To export the files, we need to create a BUILD.bazel file in the certs folder that contains the following:

```
exports_files([
  "server_cert.pem",
  "server_key.pem",
  "ca_cert.pem"
])
```

Then, in the server BUILD file, we can now add a dependency on server_cert and server_key like so (in server/BUILD.bazel):

```
go_library(
  name = "server_lib",
  //...
  data = [
    "//certs:server_cert.pem",
    "//certs:server_key.pem",
  ],
  //...
)
```

And finally, we can add the dependency to ca_cert in the client like so (in client/BUILD.bazel):

```
go_library(
  name = "client_lib",
```

```
//...
data = [
  "//certs:ca_cert.pem",
],
//...
)
```

You should now be able to run the server and the client correctly with Bazel as we showed in the previous chapters.

To conclude, we saw that we need to have a server certificate and a server key file to create a connection on the server side, and we need to have a CA certificate on the client side. We also worked with self-signed certificates but in production, these certificates should be generated for us. And finally, we saw how to create `ServerOption` and `DialOption` to enable TLS in gRPC.

Distributing requests with load balancing

Load balancing in general is a complex topic. There are many ways of implementing it. gRPC provides, by default, client-side load balancing. This is a less popular choice than look-aside or proxy load-balancing because it involves "knowing" all the servers' addresses and having complex logic in the client, but it has the advantage of directly talking to the servers and thus enables lower-latency communication. If you want to know more about how to choose the correct load balancing for your use case, check this documentation: `https://grpc.io/blog/grpc-load-balancing/`.

To see the power of client-side load balancing, we will deploy three instances of our server to Kubernetes and let the client balance the load across them. I created the Docker images beforehand so that we do not have to go through all of that here. If you are interested in checking the Docker files, you can see them both in the `server` and `client` folders. There are extensively documented. Furthermore, I uploaded the images on Docker Hub so that we can pull them easily (`https://hub.docker.com/r/clementjean/grpc-go-packt-book/tags`).

Before deploying the server and client, let us see what we need to change in terms of code. On the server side, we will simply print every request we receive. This is done with an interceptor that looks like the following (in `server/interceptors.go`):

```
func unaryLogInterceptor(ctx context.Context, req interface{}, info
*grpc.UnaryServerInfo, handler grpc.UnaryHandler) (interface{}, error)
{
  log.Println(info.FullMethod, "called")
  return handler(ctx, req)
}

func streamLogInterceptor(srv interface{}, ss grpc.ServerStream, info
  *grpc.StreamServerInfo, handler grpc.StreamHandler) error {
  log.Println(info.FullMethod, "called")
```

```
    return handler(srv, ss)
}
```

This simply prints which method has been called and continues the execution.

After that, these interceptors need to be registered in an Interceptor Chain. This is because we already have our authentication Interceptor, and gRPC accepts only one call of `grpc.UnaryInterceptor` and `grpc.StreamInterceptor`. We can now merge, in `server/main.go`, two Interceptors of the same type (unary or stream) like so:

```
opts := []grpc.ServerOption{
    //...
    grpc.ChainUnaryInterceptor(unaryAuthInterceptor,
    unaryLogInterceptor),
    grpc.ChainStreamInterceptor(streamAuthInterceptor,
    streamLogInterceptor),
}
```

That is all for the server side. Let us now focus on the client. We are going to add a `DialOption` with the `grpc.WithDefaultServiceConfig` function. This takes a JSON string as a parameter, which represents a global client configuration for the service and its methods. If you are interested in diving into the configuration, you can check the following documentation: `https://github.com/grpc/grpc/blob/master/doc/service_config.md`.

For us, the configuration will be simple; we will simply say that our client should use the `round_robin` load-balancing policy. The default policy is called `pick_first`. This is saying that the client will try to connect to all the available addresses (resolved by DNS), and once it can connect to one, it will send all the requests to that address. `round_robin` is different. It will try to connect to all the addresses available. And then, it will forward requests to each server in turn.

To set up the `round_robin` balancing, we just need to add one `DialOption` in `client/main.go`, like so:

```
opts := []grpc.DialOption{
    //...
    grpc.WithDefaultServiceConfig(`{"loadBalancingConfig":
    [{"round_robin":{}}]}`),
}
```

Finally, one last thing to note is that the load balancing only works with the DNS scheme. This means that we will change the way we run our client. Before, we had the following:

```
$ go run ./client 0.0.0.0:50051
```

Now, we will need to prepend the `dns:///` scheme, like so:

```
$ go run ./client dns:///$HOSTNAME:50051
```

Now, we are ready to talk about deploying our application. Let us start deploying the server. The first thing that we are going to need is a headless service. This is done by setting `ClusterIP` to `None`, which allows the client to find all the server instances through DNS. Each of the server instances will have its own DNS A record, which indicates the IP of the instance. On top of that, we are going to expose port `50051` to our server and make the selector equal to `todo-server` so that all the Pods with that selector will be exposed.

Right now, in `k8s/server.yaml`, we have the following:

```yaml
apiVersion: v1
kind: Service
metadata:
  name: todo-server
spec:
  clusterIP: None
ports:
  - name: grpc
    port: 50051
  selector:
    app: todo-server
```

After that, we are going to create a Deployment of 3 instances. We are going to make sure that these Deployments have the right label for the service to find them and we are going to expose port `50051`.

We can now add the following after the service:

```yaml
---

apiVersion: apps/v1
kind: Deployment
metadata:
  name: todo-server
  labels:
    app: todo-server
spec:
  replicas: 3
  selector:
    matchLabels:
      app: todo-server
  template:
    metadata:
      labels:
        app: todo-server
    spec:
      containers:
```

```
        - name: todo-server
          image: clementjean/grpc-go-packt-book:server
          ports:
          - name: grpc
            containerPort: 50051
```

We can now deploy the server instances by using the following command:

```
$ kubectl apply -f k8s/server.yaml
```

And a little bit later, we should be have the following Pods (the names might be different):

```
$ kubectl get pods
NAME                            READY    STATUS
todo-server-85cf594fb6-tkqm9    1/1      Running
todo-server-85cf594fb6-vff6q    1/1      Running
todo-server-85cf594fb6-w4s61    1/1      Running
```

Next, we need to create a Pod for the client. Normally, if the client is not a microservice, we would not have to deploy it in Kubernetes. However, since our client is a simple Go app, it would be easier to deploy it in a container to talk to our server instances.

In k8s/client.yaml, we have the following simple Pod:

```
apiVersion: v1
kind: Pod
metadata:
  name: todo-client
spec:
  containers:
  - name: todo-client
    image: clementjean/grpc-go-packt-book:client
  restartPolicy: Never
```

We can now run the client by using the following command:

```
$ kubectl apply -f k8s/client.yaml
```

And after a few seconds, we should get a similar output (or error instead of completed):

```
$ kubectl get pods
NAME           READY    STATUS
todo-client    0/1      Completed
```

Now, the most important thing is to see the actual effects of load-balancing. To do that, we will take each server name and execute a kubectl logs command for each:

```
$ kubectl logs todo-server-85cf594fb6-tkqm9
listening at 0.0.0.0:50051
/todo.v2.TodoService/UpdateTasks called
/todo.v2.TodoService/ListTasks called

$ kubectl logs todo-server-85cf594fb6-vff6q
listening at 0.0.0.0:50051
/todo.v2.TodoService/DeleteTasks called

$ kubectl logs todo-server-85cf594fb6-w4s61
listening at 0.0.0.0:50051
/todo.v2.TodoService/AddTask called
/todo.v2.TodoService/AddTask called
/todo.v2.TodoService/AddTask called
/todo.v2.TodoService/ListTasks called
/todo.v2.TodoService/ListTasks called
```

Now, you might have different results, but you should be able to see that the load was distributed across the different instances. One more thing to note is that, as we are not using a real database, the logs for the `todo-client` should not be correct. This is because we might have a `Task` on server 1 and ask to list the `Task` for server 2, which does not know about the `Task` we want. In production, we would use a real database, and that should not happen.

To conclude, we saw that the default load-balancing policy is `pick_first`, which attempts to connect to all the available addresses, in order, until it finds one that is reachable and sends all the requests to it. Then, we used a `round_robin` load-balancing policy, which sends requests to each of the servers in turn. And finally, we saw that setting up client-side load-balancing is simple in terms of gRPC code. All the rest of the configuration is mostly some DevOps work.

Summary

In this chapter, we saw the key features that we get out of the box when using gRPC. We saw that we return errors with error codes and messages. There are way fewer error codes in gRPC than in HTTP, which makes them less ambiguous.

After that, we saw that we can use the context to make a call cancelable and specify deadlines. These features are important for making reliable calls and making sure that if something goes wrong on the server side before returning, our client is not waiting indefinitely.

With context and interceptors, we also saw that we can send metadata and use them to validate requests. In our case, we checked for an authentication token every time a request was made. On the client side, we saw that interceptors can automatically add the metadata for us. This is especially useful for metadata that is shared across services and/or endpoints.

Then, we saw how we can encrypt communication over the network. We used TLS, as this is the most common way to do so. We saw that, once we have our certificates, we can simply create a `ServerOption` and a `DialOption` to let the server and client know how to understand each other.

After that, we saw how we can compress payloads. And most importantly, we saw when this might be useful and when it is not.

And finally, we used client-side load-balancing with a `round_robin` policy to distribute requests across different instances of our server.

In the next chapter, we will see more essential features of the kind we saw in this chapter. We are going to introduce the concept of middleware and see how to use different kinds of middleware to make our APIs more solid.

Quiz

1. What is the context used for?

    A. Passing metadata between the client and server

    B. Making calls cancelable

    C. Specifying a timeout

    D. All of the above

2. What is an interceptor used for?

    A. Sharing logic across endpoints

    B. Intercepting malicious data

3. What is the potential problem of using compression in gRPC?

    A. There is no problem

    B. There is the possibility that the payload gets corrupted

    C. There is the possibility that the payload gets bigger

Answers

1. D

2. A

3. C

Challenges

- Implement more errors on the server side. An example might be handling the errors coming out of `updateTask` and `deleteTasks`, which are talking to the database.

- As deadlines can save time and resources, it is important to specify them. Make all calls to our client have a deadline of 200 milliseconds.

- Create a client-side Interceptor that logs the requests the client sends.

8
More Essential Features

We saw previously that gRPC gives us a lot of important out-of-the-box features that make our job simpler. In this chapter, we are going to delve deeper into some of the important features that are not included in gRPC but are provided by the community. They generally build on top of the gRPC features to provide more convenience. They also provide a way to implement the most common practices to protect and optimize your APIs.

In this chapter, we are going to cover the following main topics:

- Validating request messages
- Creating a middleware
- Authenticating requests
- Tracing API calls
- Applying rate limiting
- Retrying on error

By the end of the chapter, we will have learned what middleware are and what they are used for. And we are going to do that by learning more about the awesome community projects called `protoc-gen-validate` and `go-grpc-middleware`.

Technical requirements

For this chapter, you will find the relevant code in the folder called `chapter8` in the accompanying GitHub repository (`https://github.com/PacktPublishing/gRPC-Go-for-Professionals/tree/main/chapter8`).

Validating requests

The first thing that we are going to do is reduce the code that checks some properties of the request messages. We are going to use the `protoc-gen-validate` plugin for `protoc`, which helps us generate validation code for certain messages. This can be useful for the use case when we check the description length and the due date of a task. We will just call a generated Validate() function and it will tell us whether the requirements for the request message are fulfilled.

The first thing that we are going to do to generate this code is to install the plugin. This is a plugin maintained by Buf and you can get it like so:

```
$ go install github.com/envoyproxy/protoc-gen-validate
```

Once we have that, we are now able to use the --validate_out option from protoc.

Now, whether we are using protoc manually or with the Buf CLI, we will need to copy the validate.proto file from the GitHub repository. This file can be found here: https://github.com/bufbuild/protoc-gen-validate/blob/main/validate/validate.proto. We will copy it into our proto folder under the validate directory:

```
proto
└── validate
    └── validate.proto
```

And now, we can import that file into other proto files and use the validation rules provided as field options.

Let us work with AddTaskRequest in proto/todo/v2/todo.proto. Right now, we have the following:

```
message AddTaskRequest {
  string description = 1;
  google.protobuf.Timestamp due_date = 2;
}
```

As we know, each time we try to add a Task, on the server side, we will check whether the description is empty or not and whether due_date is greater than time.Now() or not.

We are now going to encode this logic into our proto file. The first thing that we need to do is import the validate.proto file. Then, we will have access to the validate.rules field option, which contains a set of rules for multiple types. We are going to work on string and Timestamp and we are going to use the min_len and gt_now fields. The first one describes the minimum length that the string should have when we call Validate and the second one tells us that the Timestamp provided should be in the future:

```
import "validate/validate.proto";

//...

message AddTaskRequest {
```

```
string description = 1 [
  (validate.rules).string.min_len = 1
];

google.protobuf.Timestamp due_date = 2 [
  (validate.rules).timestamp.gt_now = true
];
}
```

Now that we have described this logic, we will need to generate code that checks that logic. Otherwise, these options are worthless. To generate this code, we are going to do it manually with protoc and then I will show you how to do it with Buf and Bazel.

As mentioned, with the plugin we can use the `--validate_out` option in `protoc`. This looks like the following:

```
$ protoc -Iproto --go_out=proto --go_opt=paths=
  source_relative --go-grpc_out=proto --go-grpc_opt=
    paths=source_relative --validate_out=
"lang=go,paths=source_relative:proto" proto/todo/v2/*.proto
```

Notice that the command is similar to what we ran in the past. We simply added the new option and told it to work on Go code and to generate code based on the proto files in the v2 folder.

And now, on top of the Protobuf and gRPC generated code, you should also have a `.pb.validate.go` file in the v2 folder. This should look like this:

```
proto/todo/v2
├── todo.pb.go
├── todo.pb.validate.go
├── todo.proto
└── todo_grpc.pb.go
```

Inside the generated file, you should be able to see the following function (among others):

```
// Validate checks the field values on Task with the rules
defined in the proto
// definition for this message. If any rules are violated,
the first error
// encountered is returned, or nil if there are no
violations.
func (m *Task) Validate() error {
  return m.validate(false)
}
```

This is the function that we are now going to use in our `AddTask` endpoint on the server side. Right now, we have the following checks:

```
func (s *server) AddTask(_ context.Context, in
*pb.AddTaskRequest) (*pb.AddTaskResponse, error) {
  if len(in.Description) == 0 {
    return nil, status.Error(
      codes.InvalidArgument,
      "expected a task description, got an empty string",
    )
  }

  if in.DueDate.AsTime().Before(time.Now().UTC()) {
    return nil, status.Error(
      codes.InvalidArgument,
      "expected a task due_date that is in the future",
    )
  }
  //...
}
```

Let us replace that with the `Validate` function. We are simply going to call the function on the `in` parameter and if it returns any error, we will return the error from the function, otherwise, we will simply continue with our execution:

```
func (s *server) AddTask(_ context.Context, in
*pb.AddTaskRequest) (*pb.AddTaskResponse, error) {
  if err := in.Validate(); err != nil {
    return nil, err
  }
  //...
}
```

It is as simple as that, and we saved ourselves from writing all the checks manually and trying to keep our error messages consistent across the different endpoints.

We can now go to the `main` client and uncomment the functions, one by one, in the error section:

```
func main() {
  //...

  fmt.Println("-------ERROR-------")
  addTask(c, "", dueDate)
  addTask(c, "not empty", time.Now().Add(-5*time.Second))
```

```
    fmt.Println("-------------------")
}
```

We should get the following error for the first `addTask`:

```
$ go run ./client 0.0.0.0:50051
-------ERROR-------
rpc error: code = Unknown desc = invalid AddTaskRequest
.Description: value length must be at least 1 runes
```

And this one for the second `addTask`:

```
$ go run ./client 0.0.0.0:50051
-------ERROR-------
rpc error: code = Unknown desc = invalid AddTaskRequest
.DueDate: value must be greater than now
```

Notice that the code error is Unknown. As of the time of writing this book, `protoc-gen-validate` does not appear to have a custom error code. This might appear in `v2` of the plugin. However, it provides us with a simple validation code and clean error messages.

Buf

Using `protoc-gen-validate` with the Buf CLI is simple. We will add some configuration in our YAML files in order to generate the code. The first thing that we need to add is the dependency on `protoc-gen-validate` in our `buf.yaml` file:

```
version: v1
#...
deps:
- buf.build/envoyproxy/protoc-gen-validate
```

This tells Buf that we need `protoc-gen-validate` during the generation process. It will later figure out how to pull the dependency by itself.

And after that, we need to configure the plugin in the `buf.gen.yaml` file:

```
version: v1
plugins:
  #...
  - plugin: buf.build/bufbuild/validate-go
    out: proto
    opt: paths=source_relative
```

These options are the same as we typed manually earlier. Now, we can generate as usual by typing the following command:

```
$ buf generate proto
```

You should now have the same three generated files that we obtained with the `protoc` command: `todo.pb.validate.go`, `todo_grpc.pb.go`, and `todo.pb.go`. Note that, in this case, we also generated code for `v1` and the `validate.proto`.

Bazel

As always, the first thing we need to do is define the dependency in our `WORKSPACE.bazel` file. We are going to fetch the `protoc-gen-validate` project from GitHub and load its relevant dependencies:

> **Important note**
>
> The code that follows references a variable called `PROTOC_GEN_VALIDATE_VERSION`. This variable is defined in the `versions.bzl` file in the `chapter8` folder. We do not include it here to keep the code independent of versions.

```
#...
git_repository(
  name = "com_envoyproxy_protoc_gen_validate",
  tag = PROTOC_GEN_VALIDATE_VERSION,
  remote = "https://github.com/bufbuild/protoc-gen-validate"
)

load("@com_envoyproxy_protoc_gen_validate//bazel:
  repositories.bzl", "pgv_dependencies")
load("@com_envoyproxy_protoc_gen_validate//
  :dependencies.bzl", "go_third_party")

pgv_dependencies()

# gazelle:repository_macro deps.bzl%go_third_party
go_third_party()
```

With that, we now need to update our dependencies in the `deps.bzl` file. We can do that by typing the following command:

```
$ bazel run //:gazelle-update-repos
```

And finally, we need to generate the code and link it to our existing todo go_library in proto/todo/v2/BUILD.bazel.

The first thing to add is the dependency on protoc-gen-validate validate.proto in v2_proto proto_library. This will allow todo.proto to import it:

```
proto_library(
  name = "v2_proto",
  #…
  deps = [
    #...
    "@com_envoyproxy_protoc_gen_validate//
      validate:validate_proto",
  ],
)
```

Then, we will replace v2_go_proto go_proto_library with pgv_go_proto_library (**pgv** stands for **protoc-gen-validate**). On top of that, we will add the dependency to the protoc-gen-validate library so that the generated code accesses any of the protoc-gen-validate internal code needed to compile:

```
load("@com_envoyproxy_protoc_gen_validate//bazel:pgv_proto_
  library.bzl", "pgv_go_proto_library")
//...
pgv_go_proto_library(
  name = "v2_go_proto",
  compilers = ["@io_bazel_rules_go//proto:go_grpc"],
  importpath = "github.com/PacktPublishing/gRPC-Go-for-Professionals/
    proto/todo/v2",
  proto = ":v2_proto",
  deps = ["@com_envoyproxy_protoc_gen_validate//
    validate:validate_go"],
)
```

And finally, in order to avoid ambiguous imports for validate/validate.proto the next time we run Gazelle, we are going to map the validate/validate.proto import (in proto/todo/v2/todo.proto) to @com_envoyproxy_protoc_gen_validate//validate:validate_proto (defined in protoc-gen-validate). At the top of the proto/todo/v2/BUILD.bazel file, we can add the following Gazelle directive:

```
# gazelle:resolve proto validate/validate.proto
  @com_envoyproxy_protoc_gen_validate//validate:
  validate_proto
```

Now that we have replaced our old v2_go_proto with the new one using pgv_go_proto_library, the code depending on this library will automatically get access to the generated Validate function.

We can try running the server:

```
$ bazel run //server:server 0.0.0.0:50051
listening at 0.0.0.0:50051
```

And run the client with the error section code uncommented:

```
$ bazel run //client:client 0.0.0.0:50051
-------ERROR-------
rpc error: code = Unknown desc = invalid AddTaskRequest
.Description: value length must be at least 1 runes
```

To conclude, we saw that we can encode validation logic in our proto files and generate validation code automatically with protoc-gen-validate. This declutters our code and provides consistent error messages across our API endpoints.

Middleware = interceptor

In the context of gRPC, a middleware is an interceptor. It lies between the code registered by the developers and the actual gRPC framework. When gRPC receives some data from the wire, it will pass this data through the middleware first, and then if it is allowed to go through, the data will arrive in the actual endpoint handler.

These middleware are generally used in order to secure the endpoints against malicious actors or enforce certain prerequisites. An example of securing the API is rate-limiting clients. This is a limit on the number of requests that a client can make in a given timeframe and this is important because it prevents a lot of attacks, such as brute force attacks, DoS and DDoS, and web-scraping. And to enforce a certain prerequisite, we already saw an example where the client needs to be authenticated before being able to call an endpoint.

Before going to see some of the middleware provided by the community, I want to remind you that we already created middleware in *Chapter 7*. We simply never referred to them as middleware. In fact, we created two and they are the following:

```
opts := []grpc.ServerOption{
  //...
  grpc.ChainUnaryInterceptor(unaryAuthInterceptor,
    unaryLogInterceptor),
  grpc.ChainStreamInterceptor(streamAuthInterceptor,
    streamLogInterceptor),
}
```

If you remember, these middleware will first check that there is an `auth_token` header that exists and that its value is `authd`. And then, if it is the case, it will log the API call on the terminal and continue with the execution of the code we wrote for the API endpoint.

So, to summarize, a middleware is an interceptor that can cut the execution short depending on some conditions and is there to secure the API endpoints.

Authenticating requests

In this section and the following, we are going to simplify the middleware that we currently have. First, we are going to start by simplifying the authentication process. We saw in the previous chapter that we can easily create an interceptor for checking an authentication token in headers. In this section, we are going to take a step further and make it even simpler.

> **Important note**
>
> gRPC supports retrying authentication of requests through an RBAC policy without a third-party library. However, the configuration is quite verbose and not very well documented. If you are interested in trying it, you can check the following example: `https://github.com/grpc/grpc-go/blob/master/examples/features/authz/README.md`.

Previously, when we wrote our interceptors, we needed to create the following function for a unary interceptor:

```
func unaryAuthInterceptor(ctx context.Context, req
  interface{}, info *grpc.UnaryServerInfo, handler
    grpc.UnaryHandler) (interface{}, error)
```

And another one like the following for the stream interceptor:

```
func streamAuthInterceptor(srv interface{}, ss
  grpc.ServerStream, info *grpc.StreamServerInfo, handler
    grpc.StreamHandler) error
```

While this gives us a lot of information on the calls, the context, and so on, this also makes our code terse, and we need to think about how to share the common business logic between the stream and unary interceptors.

With the middleware that we are going to add, we will simply focus on our logic, and we will be able to register the interceptor as easily as before in our gRPC server. This middleware is the auth middleware in the GitHub repository called `go-grpc-middleware` (`https://github.com/grpc-ecosystem/go-grpc-middleware`). It will let us get rid of the complicated authenticator function definitions that we add for the interceptors and will let us register directly using our `validateAuthToken` function in a predefined interceptor.

To get started, we are going to fetch the dependency in our `server` folder:

```
$ go get github.com/grpc-ecosystem/go-grpc-middleware/v2/
  interceptors/auth
```

Then, we are going to remove the `unaryAuthInterceptor` and `streamAuthInterceptor` in the `server/interceptors.go` file. We do not need them anymore since the new auth middleware will take care of everything for us.

And finally, we are going to go to the `server/main.go`, where we are going to replace the old interceptor with `auth.UnaryServerInterceptor` and `auth.StreamServerInterceptor`. These two interceptors take an `AuthFunc`, which basically represents the authentication logic. In our case, we will pass to them our `validateAuthToken`.

The `AuthFunc` type looks like this:

```
type AuthFunc func(ctx context.Context) (context.Context,
error)
```

So, we need to change the `validateAuthToken` a little bit to return a context and/or an error. Our new function will look like this:

```
func validateAuthToken(ctx context.Context)
  (context.Context, error) {
  md, ok := metadata.FromIncomingContext(ctx)

  if !ok {
    return nil, status.Errorf(/*...*/)
  }

  if t, ok := md["auth_token"]; ok {
    switch {
    case len(t) != 1:
      return nil, status.Errorf(/*...*/)
    case t[0] != "authd":
      return nil, status.Errorf(/*...*/)
    }
  } else {
    return nil, status.Errorf(/*...*/)
  }

  return ctx, nil
}
```

This lets us register the `validateAuthToken` in the gRPC server. Our new main function will now be like the following:

```go
import (
  //...
  "github.com/grpc-ecosystem/go-grpc-middleware/v2/
    interceptors/auth"
)

func main() {
  //...
  opts := []grpc.ServerOption{
    //...
    grpc.ChainUnaryInterceptor
      (auth.UnaryServerInterceptor(validateAuthToken),
        unaryLogInterceptor),
    grpc.ChainStreamInterceptor(auth
      .StreamServerInterceptor(validateAuthToken),
        streamLogInterceptor),
  }
  //...
}
```

Now, we should be able to run the server:

```
$ go run ./server 0.0.0.0:50051
listening at 0.0.0.0:50051
```

We also run the client:

```
$ go run ./client 0.0.0.0:50051
```

We don't get any errors and it works similarly to before. However, to test that the middleware is working properly, we can temporarily modify our interceptor on the client side to add a wrong authentication header (`client/interceptors.go`):

```go
const authTokenValue string = "notauthd"
```

And if we rerun the client, we should get the following error:

```
$ go run ./client 0.0.0.0:50051
--------ADD--------
rpc error: code = Unauthenticated desc = incorrect
  auth_token
```

This proves that our middleware is working as expected and that we can rely on only `validAuthToken` to do the authentication checking.

Bazel

In order to run that with Bazel, we will need to update our dependencies and link the new dependency to the `server_lib` target in `server/BUILD.bazel`. So, we first run the `gazelle-update-repos` command, which will fetch the `go-grpc-middleware` dependency:

```
$ bazel run //:gazelle-update-repos
```

And once we have that, we can now let the `gazelle` command include the `go-grpc-middleware` dependency to the target:

```
$ bazel run //:gazelle
```

Finally, we will be able to run our server:

```
$ bazel run //server:server 0.0.0.0:50051
listening at 0.0.0.0:50051
```

And the client with the wrong `auth` token should give the following message:

```
$ bazel run //client:client 0.0.0.0:50051
--------ADD--------
rpc error: code = Unauthenticated desc = incorrect
  auth_token
```

To conclude, in this section, we saw that we can simplify the authenticator interceptors by using the `go-grpc-middleware` package. It lets us focus on the actual logic and not on how to write an interceptor that can be registered with gRPC.

Logging API calls

In this section, let us simplify the log interceptor. This will be like what we did in the previous section, but we are going to use another middleware: the logging middleware.

While this middleware integrates with a lot of different loggers, we are going to use it with the default `log` package in Golang. It will then appear easy to integrate with your favorite logger.

> **Important note**
>
> The next command is only needed if you did not get the previous dependency on `go-grpc-middleware`. If you followed section by section, you should not need it.

To get started, let us get the dependency on the middleware. In the `server` folder, we are going to run the following command:

```
$ go get github.com/grpc-ecosystem/go-grpc-middleware/v2/
  interceptors/logging
```

Now, we can start creating our logger. We are going to create it by defining a function that returns a `loggerFunc`. This is a function that has the following signature:

```
func(ctx context.Context, lvl logging.Level, msg string,
  fields ...any)
```

We already know what the context is, but all the rest is specific to the logger. The level is a logging level such as `Debug`, `Info`, `Warning`, or `Error`. This is generally used in order to filter the logs depending on the level of severity. Then, the message is simply a message generated by the logger middleware such as `":started call"` or `":finished call"`. This helps us understand the context of the log. And finally, the fields are all the other information that we need to print a useful log. In our case, we are going to use the service name and the method name. This will let us create logs like the following:

```
INFO :started call todo.v2.TodoService UpdateTasks
```

A thing that is not easy to wrap your mind around is the `fields` parameter. This is because this is presented as a `vararg` of any. In reality, we can transform that into a map in order to get a specific field name such as `grpc.service`, `grpc.method`, ... To do that, we can simply write the following:

```
f := make(map[string]any, len(fields)/2)
i := logging.Fields(fields).Iterator()

for i.Next() {
  k, v := i.At()
  f[k] = v
}
```

Notice that we are creating a map of length `len(fields)/2`. This is because in the `fields` parameter, the name of the fields and their value are interleaved. An example is the following:

```
grpc.service todo.v2.TodoService    grpc.method ListTasks
```

You can print the fields and see the whole thing yourself by expanding the `vararg`:

```
log.Println(fields...)
```

Now that we have this knowledge, we can proceed to write the logger. We will create a function called `logCalls`, which takes a `log.Logger` (from the `golang` standard library) as a param and will return a `logging.Logger` (from the logging middleware). The logic of the logger will be to check the log level, prepend the level of the message to it, and then we will append the service name and method name to the whole message:

```go
import (
  //...
  "github.com/grpc-ecosystem/go-grpc-middleware/v2/interceptors/
logging"
)

const grpcService = "grpc.service"
const grpcMethod = "grpc.method"

func logCalls(l *log.Logger) logging.Logger {
  return logging.LoggerFunc(func(_ context.Context, lvl
   logging.Level, msg string, fields ...any) {
    f := make(map[string]any, len(fields)/2)
    i := logging.Fields(fields).Iterator()

    for i.Next() {
      k, v := i.At()
      f[k] = v
    }

    switch lvl {
    case logging.LevelDebug:
      msg = fmt.Sprintf("DEBUG :%v", msg)
    case logging.LevelInfo:
      msg = fmt.Sprintf("INFO :%v", msg)
    case logging.LevelWarn:
      msg = fmt.Sprintf("WARN :%v", msg)
    case logging.LevelError:
      msg = fmt.Sprintf("ERROR :%v", msg)
    default:
      panic(fmt.Sprintf("unknown level %v", lvl))
    }

    l.Println(msg, f[grpcService], f[grpcMethod])
  })
}
```

Now, while this method will always be accurate because we can retrieve the keys in the map that was built, this means that we need to build a map each time this interceptor is called. This is not really efficient. I wanted to show you the full-blown example before showing you the efficient one so that you understand how to use the fields parameter.

To be more efficient, we can take advantage of the fact that our service and method are always situated at index 5 and 7, respectively. So, we are going to remove the map creation part, we are going to replace `grpcService` and `grpcMethod` with 5 and 7, and we are going to access the 5th and 7th element of fields:

```
const grpcService = 5
const grpcMethod = 7

func logCalls(l *log.Logger) logging.Logger {
  return logging.LoggerFunc(func(_ context.Context, lvl
    logging.Level, msg string, fields ...any) {
    // ...
    l.Println(msg, fields[grpcService], fields[grpcMethod])
  })
}
```

This is much more efficient. Now, one thing that is worth mentioning is that this is less safe. We are assuming that all the fields that we receive will always contain the `service` and `method` at the same index and that our `fields` array is large enough. We can safely assume that now, at the time of writing, because these are common fields that are added in this order all the time. However, if the library changes, you might try to do an out-of-bounds access or get a different piece of information. Be aware of that.

The last thing that we need to do is register the interceptor. This is similar to what we did with the authentication interceptor, but the main difference is that now we need to create a logger and pass it to the `logCalls` function. We are going to use a golang `log.Logger`, which prints the date and time before the message. And finally, we are going to pass the result of `logCalls` to `logging.UnaryServerInterceptor` and `logging.StreamServerInterceptor`:

```
import (
  //...
  "github.com/grpc-ecosystem/go-grpc-middleware/v2/
    interceptors/logging"
)

//...

func main() {
  //...
  logger := log.New(os.Stderr, "", log.Ldate|log.Ltime)
  opts := []grpc.ServerOption{
    //...
    grpc.ChainUnaryInterceptor(
      //...
```

```
        logging.UnaryServerInterceptor(logCalls(logger)),
    ),
    grpc.ChainStreamInterceptor(
      //...
      logging.StreamServerInterceptor(logCalls(logger)),
    ),
  }
  //...
}
```

After that, we can now run our server:

```
$ go run ./server 0.0.0.0:50051
listening at 0.0.0.0:50051
```

> **Note**
>
> Before running the client make sure that you replaced the value of authTokenValue to authd in the client/interceptors.go file

And then run our client:

```
$ go run ./client 0.0.0.0:50051
```

And if we check the terminal in which the server is running, we should have a bunch of messages like the following:

```
INFO :started call todo.v2.TodoService ListTasks
INFO :finished call todo.v2.TodoService ListTasks
```

To conclude, we saw that, similarly to the authentication middleware, we can simply add a logger to our gRPC server. We also saw that we can access more information than just the service name and method name by transforming the fields varargs into a map. And finally, we saw that some of the fields are always at the same place in vararg, so instead of generating the map for each call, we can directly access the information by index.

Tracing API calls

On top of logging, which simply describes the events in a developer-friendly manner, you might need to get metrics that can get aggregated by dashboard tools. These metrics might include requests per second, distribution of status (Ok, Internal, and so on), and many others. In this section, we are going to instrument our code with OpenTelemetry and Prometheus so that tools such as Grafana can be used to create dashboards.

The first thing to understand is that we are going to run an HTTP server for Prometheus metrics. Prometheus exposes the metrics to external tools on the /metrics route so that the tools wanting to query the data get a sense of all the kinds of metrics available.

So, to create such a server, we are going to get a dependency on Prometheus' Go library. We are going to do that by getting the dependency on the go-grpc-middleware/providers/prometheus. The Prometheus Go library is a transitive dependency of this one and we still need to be able to register some more interceptors that are defined in the Prometheus provider:

```
$ go get github.com/grpc-ecosystem/go-grpc-middleware/
  providers/prometheus
```

Now, we can create an HTTP server that will later be used to expose the /metrics route. We are going to create a function called newMetricsServer, which takes the address of where the server is running:

> **Important note**
>
> The code that follows explains every part of the server/main.go file. Displaying the full file here would be overwhelming. Thus, we will walk through all of its code and you will be able to see the imports and the overall structure in the main.go. Note that, in order to better explain, We are only adding certain elements later in the section. If you see a part of the code that is not presented yet, read on and you will get an explanation for the piece of code you are looking at.

```go
func newMetricsServer(httpAddr string) *http.Server {
  httpSrv := &http.Server{Addr: httpAddr}
  m := http.NewServeMux()
  httpSrv.Handler = m
  return httpSrv
}
```

Now that we have our HTTP server, we will refactor our main and separate the creation of the gRPC server into another function:

```go
func newGrpcServer(lis net.Listener) (*grpc.Server, error) {
  creds, err := credentials.NewServerTLSFromFile
    ("./certs/server_cert.pem", "./certs/server_key.pem")
  if err != nil {
    return nil, err
  }

  logger := log.New(os.Stderr, "", log.Ldate|log.Ltime)

  opts := []grpc.ServerOption{
    //...
```

```
    }
    s := grpc.NewServer(opts...)

    pb.RegisterTodoServiceServer(s, &server{
      d: New(),
    })
    return s, nil
}
```

Nothing really changed; we only separated the creation into a function so that we can run two servers in parallel later. Before working on that though, our `main` should also include two addresses as parameters. The first one is for the gRPC server and the other is for the HTTP server:

```
func main() {
    args := os.Args[1:]

    if len(args) != 2 {
      log.Fatalln("usage: server [GRPC_IP_ADDR]
        [METRICS_IP_ADDR]")
    }

    grpcAddr := args[0]
    httpAddr := args[1]
}
```

Now, we can deal with running two servers at once. We are going to use the `errgroup` (`https://pkg.go.dev/golang.org/x/sync/errgroup`) package. It lets us add multiple goroutines to a group and wait on them.

The first thing that we need is to create a context for the group. We are going to create a cancellable one so that later we can release the servers' resources:

```
ctx := context.Background()
ctx, cancel := context.WithCancel(ctx)
defer cancel()
```

Next, we can start handling the `SIGTERM` signal. This is because when we want to exit both servers, we will press *Ctrl + C*. This will send the `SIGTERM` signal, and we expect the servers to be closed gracefully. To handle that, we are going to make a channel that will be released when the `SIGTERM` signal is received:

```
quit := make(chan os.Signal, 1)
signal.Notify(quit, os.Interrupt, syscall.SIGTERM)
defer signal.Stop(quit)
```

After that, we can now create our group of two servers. We will first create the group from the cancellable context that we created. And then, we will use the Go(func() error) function to add goroutines to that group. The first goroutine will handle the serving of the gRPC server and the second goroutine will handle the HTTP one:

```go
lis, err := net.Listen("tcp", grpcAddr)

if err != nil {
  log.Fatalf("unexpected error: %v", err)
}
g, ctx := errgroup.WithContext(ctx)
grpcServer, err := newGrpcServer(lis)

if err != nil {
  log.Fatalf("unexpected error: %v", err)
}

g.Go(func() error {
  log.Printf("gRPC server listening at %s\n", grpcAddr)
  if err := grpcServer.Serve(lis); err != nil {
    log.Printf("failed to gRPC server: %v\n", err)
    return err
  }
  log.Println("gRPC server shutdown")
  return nil
})


metricsServer := newMetricsServer(httpAddr)
g.Go(func() error {
  log.Printf("metrics server listening at %s\n", httpAddr)
  if err := metricsServer.ListenAndServe(); err != nil &&
    err != http.ErrServerClosed {
    log.Printf("failed to serve metrics: %v\n", err)
    return err
  }
  log.Println("metrics server shutdown")
  return nil
})
```

Now that we have our group, we can wait on the context to be done or on the quit channel to receive an event:

```go
select {
  case <-quit:
```

```
      break
   case <-ctx.Done():
      break
}
```

Once one of these events is received, we are going to initiate the release of resources by making sure that the context is finished (call the `cancel` function), and finally, we can wait on the group to finish all the goroutines that we registered:

```
cancel()

timeoutCtx, timeoutCancel := context.WithTimeout(
   context.Background(),
   10*time.Second,
)
defer timeoutCancel()

log.Println("shutting down servers, please wait...")

grpcServer.GracefulStop()
metricsServer.Shutdown(timeoutCtx)

if err := g.Wait(); err != nil {
   log.Fatal(err)
}
```

Finally, as this is the ultimate goal of this section, we need to add tracing capabilities. The metrics server will expose the metrics route and the gRPC server will collect the metrics and add them to the Prometheus registry. This registry is a collection of collectors. We register one or more collectors to it, then the registry will collect the different metrics available, and finally, it will expose these metrics.

Before creating the registry, we will first create a collector with the `NewServerMetrics` function provided in the `go-grpc-middleware/providers/prometheus` package. Then, we are going to actually create the registry. And finally, we are going to register the collector:

```
srvMetrics := grpcprom.NewServerMetrics(
   grpcprom.WithServerHandlingTimeHistogram(
      grpcprom.WithHistogramBuckets([]float64{0.001, 0.01,
         0.1, 0.3, 0.6, 1, 3, 6, 9, 20, 30, 60, 90, 120}),
   ),
)
reg := prometheus.NewRegistry()
reg.MustRegister(srvMetrics)
```

Notice that we passed an option to the `NewServerMetrics`. This option will let us get buckets into which the calls will be placed depending on their latency. This basically tells us how many requests were served in under 0.001 seconds, 0.01 seconds, and so on.

Finally, we are going to pass the registry to the HTTP server so that it knows what metrics are available, and we are going to pass the collector to our gRPC server so that it can push metrics to it:

```go
func newMetricsServer(httpAddr string, reg
  *prometheus.Registry) *http.Server {
  //...
  m.Handle("/metrics", promhttp.HandlerFor(reg,
    promhttp.HandlerOpts{}))
  //...
  return httpSrv
}

func newGrpcServer(lis net.Listener, srvMetrics
  *grpcprom.ServerMetrics) (*grpc.Server, error) {
  //...
  opts := []grpc.ServerOption{
    //...
    grpc.ChainUnaryInterceptor(
      otelgrpc.UnaryServerInterceptor(),
      srvMetrics.UnaryServerInterceptor(),
      //...
    ),
    grpc.ChainStreamInterceptor(
      otelgrpc.StreamServerInterceptor(),
      srvMetrics.StreamServerInterceptor(),
      //...
    ),
  }
  //...
}
func main() {
  //...
  grpcServer, err := newGrpcServer(lis, srvMetrics)
  //...
  metricsServer := newMetricsServer(httpAddr, reg)
  //...
}
```

Notice that we are now using opentelemetry (otelgrpc). This is a tool that lets us generate all the metrics from our gRPC server automatically. Then, Prometheus will pick those with the collector (srvMetrics). And finally, the HTTP server will be able to expose these metrics.

To get OpenTelemetry for gRPC, we simply need to get the dependency:

```
$ go get go.opentelemetry.io/contrib/instrumentation/
  google.golang.org/grpc/otelgrpc
```

We should now be able to run our server:

```
$ go run ./server 0.0.0.0:50051 0.0.0.0:50052
metrics server listening at 0.0.0.0:50052
gRPC server listening at 0.0.0.0:50051
```

Then, we can run our client against the 0.0.0.0:50051 address:

```
$ go run ./client 0.0.0.0:50051
```

And after the client calls are all served, we can look at the metrics like so:

```
$ curl http://localhost:50052/metrics
```

You should now have logs that look like this (simplified to only show AddTask):

```
grpc_server_handled_total{grpc_code="OK",grpc_method="
  AddTask",grpc_service="todo.v2.TodoService",grpc_type=
    "unary"} 3
grpc_server_handling_seconds_bucket{grpc_method="AddTask",
  grpc_service="todo.v2.TodoService",grpc_type="unary",le=
    "0.001"} 3
grpc_server_handling_seconds_sum{grpc_method="AddTask",grpc
  _service="todo.v2.TodoService",grpc_type="unary"}
    0.000119291
grpc_server_msg_received_total{grpc_method="AddTask",grpc_
  service="todo.v2.TodoService",grpc_type="unary"} 3
grpc_server_msg_sent_total{grpc_method="AddTask",grpc_
  service="todo.v2.TodoService",grpc_type="unary"} 3
grpc_server_started_total{grpc_method="AddTask",grpc_
  service="todo.v2.TodoService",grpc_type="unary"} 3
```

These metrics mean that the server received three AddTask, handled them all in under 0.001 seconds (total: 0.000119291), and returned three responses to the client.

There is obviously a lot more to do with these metrics. However, that would probably be a book in itself. If you are interested in this area, I would encourage you to look at how to integrate Prometheus with a tool such as Grafana to create dashboards representing these metrics in a more human-readable manner.

Bazel

We need to update the dependencies in order to get Prometheus and OpenTelemetry to work. To do that, we are going to run `gazelle-update-repos`:

```
$ bazel run //:gazelle-update-repos
```

And then, we are going to run `gazelle` in order to automatically link the dependencies to our code:

```
$ bazel run //:gazelle
```

Finally, we can now run our server:

```
$ bazel run //:server:server 0.0.0.0:50051 0.0.0.0:50052
metrics server listening at 0.0.0.0:50052
gRPC server listening at 0.0.0.0:50051
```

And then run our client:

```
$ bazel run //client:client 0.0.0.0:50051
```

In conclusion, we saw how we can get metrics out of our gRPC server by using OpenTelemetry and Prometheus. We did that by creating a second server exporting metrics on the `/metrics` route, and through the use of a Prometheus registry and collector, we exchanged metrics from the gRPC server to the HTTP server.

Securing APIs with rate limiting

For the last interceptor we are going to add to the server, we are going to use a rate limiter. More precisely, we are going to use the implementation of a token bucket rate limiter, which is provided by the `golang.org/x/time/rate` package. In this section, we are not going to delve deeply into what rate limiters are or how to build one – that is out of the scope of this book, however, you will see how you can use a rate limiter (a readily implemented or custom one) in the context of gRPC.

The first thing that we need to do is get the dependency on the rate limiter:

```
$ go get golang.org/x/time/rate
```

> **Important note**
> The next command is only needed if you did not get the previous dependency on `go-grpc-middleware`. If you followed section by section, you should not need it.

The we get the dependency for the interceptor:

```
$ go get github.com/grpc-ecosystem/go-grpc-middleware/v2/
  interceptors/ratelimit
```

Now, we are going to create a file called `limit.go`, which will contain our logic and the wrapper around `rate.Limiter`. We create such a wrapper because the interceptor that we are going to use later requires the limiter to implement a function called `Limit`, taking a context as a parameter and `rate.Limiter` does not have such a function:

```go
package main

import (
  "context"
  "fmt"

  "golang.org/x/time/rate"
)

type simpleLimiter struct {
  limiter *rate.Limiter
}

func (l *simpleLimiter) Limit(_ context.Context) error {
  if !l.limiter.Allow() {
    return fmt.Errorf("reached Rate-Limiting %v", l
      .limiter.Limit())
  }

  return nil
}
```

Notice that we simply check that the rate limiter allows (or not) the call to pass. If it does not, we return an error, otherwise, we return `nil`.

The last thing to do is to register `simpleLimiter` in an interceptor. We are going to create an instance of type `rate.Limiter` with 2 tokens per second (referred to as r) and a burst size of 4 (referred to as b). If you are unclear on what those parameters are, we recommend you read the documentation for Limiter (`https://pkg.go.dev/golang.org/x/time/rate#Limiter`):

```go
import (
  //...
  "github.com/grpc-ecosystem/go-grpc-middleware/v2/
    interceptors/ratelimit"
)
```

```
func newGrpcServer(lis net.Listener, srvMetrics
  *grpcprom.ServerMetrics) (*grpc.Server, error) {
  //...
  limiter := &simpleLimiter{
    limiter: rate.NewLimiter(2, 4),
  }
  opts := []grpc.ServerOption{
    //...
    grpc.ChainUnaryInterceptor(
      ratelimit.UnaryServerInterceptor(limiter),
      //...
    ),
    grpc.ChainStreamInterceptor(
      ratelimit.StreamServerInterceptor(limiter),
      //...
    ),
  }
  //...
}
```

That is all. We now have rate limiting enabled for our API. We can now run our server:

```
$ go run ./server 0.0.0.0:50051 0.0.0.0:50052
metrics server listening at 0.0.0.0:50052
gRPC server listening at 0.0.0.0:50051
```

And then we can try to execute more than two calls per second. This should not be hard. In fact, you normally can run your client once and it should fail. But in order to be sure that it fails, run the client multiple times. On Linux and Mac, you can run the following:

```
$ for i in {1..10}; do go run ./client 0.0.0.0:50051; done
```

And on Windows (PowerShell), you can run this:

```
$ foreach ($item in 1..10) { go run ./client 0.0.0.0:50051 }
```

You should see some queries returning responses and then, quickly, you should be able to see the following message:

```
rpc error: code = ResourceExhausted desc =
  /todo.v2.TodoService/UpdateTasks is rejected by
    grpc_ratelimit middleware, please retry later. reached
      Rate-Limiting 2
```

Obviously, our rate is very low, and it is not practical in production. We chose such a low rate in order to show you how to rate limit. In production, you will have business-specific requirements that need to be followed. You will have to adapt the code we have shown to match these requirements.

Bazel

In order to run this example with Bazel, we will need to update the repos and run Gazelle to import the new dependency (`golang.org/x/time/rate`) to our library:

```
$ bazel run //:gazelle-update-repos
$ bazel run //:gazelle
```

After that, you should be able to run the server like so:

```
$ bazel run //server:server 0.0.0.0:50051 0.0.0.0:50052
```

To conclude, we saw that we can integrate a rate limiter in our gRPC server. The `go-grpc-middleware` interceptor for rate limiting makes it easy to add a readily available implementation or a custom one.

Retrying calls

As of now, we have worked only on the server side. Let us now see an important feature on the client side. This feature is the retrying of calls that failed depending on the status code. This might be interesting for use cases where the network is unreliable. If we get an `Unavailable` error code, we will retry with an exponentially bigger wait time. This is because we do not want to retry too often and overload the network.

> **Important note**
>
> gRPC supports retries without the need for a third-party library. However, the configuration is quite verbose and not very well documented. If you are interested in trying it, you can check the following example: `https://github.com/grpc/grpc-go/blob/master/examples/features/retry/README.md`.

Let us get the dependency that we need (`client` folder):

```
$ go get github.com/grpc-ecosystem/go-grpc-middleware/v2/
    interceptors/retry
```

Then, we can define some options for the retry. We will define how many times and on which error code we want to retry. We want to retry 3 times, with exponential backoff (starting at 100 ms), with the error code being `Unavailable`:

```
retryOpts := []retry.CallOption{
    retry.WithMax(3),
```

```
retry.WithBackoff(retry.BackoffExponential(100 *
  time.Millisecond)),
retry.WithCodes(codes.Unavailable),
}
```

And then, we simply pass these options to the interceptors provided by the `retry` package:

```
import (
  //...
  "github.com/grpc-ecosystem/go-grpc-middleware/v2/
    interceptors/retry"
)

func main() {
  //...
  retryOpts := []retry.CallOption{
    //...
  }
  opts := []grpc.DialOption{
    //...
    grpc.WithChainUnaryInterceptor(
      retry.UnaryClientInterceptor(retryOpts...),
      //...
    ),
    grpc.WithChainStreamInterceptor(
      retry.StreamClientInterceptor(retryOpts...),
      //...
    ),
    //...
  }
  //...
}
```

> **Important note**
>
> Retrying is not available for client streaming. If you attempt to retry on such an RPC endpoint, you will get the following error: `rpc error: code = Unimplemented desc = grpc_retry: cannot retry on ClientStreams, set grpc_retry. Disable()`. As such, it is a bit risky to add the `retry.StreamClientInterceptor` as presented. We just wanted to show you that some streaming could also be retried.

Once we have that, we now have a problem. Our API is running locally and there is little chance that we will get an `Unavailable` error. So, for the sake of testing and demonstration, we are going to make our `AddTask` directly return such an error temporarily. In `server/impl.go`, we can comment on the rest of the function and add the following:

```
func (s *server) AddTask(_ context.Context, in
  *pb.AddTaskRequest) (*pb.AddTaskResponse, error) {
  return nil, status.Errorf(
    codes.Unavailable,
    "unexpected error: %s",
    "unavailable",
  )
}
```

And now, we run our server:

```
$ go run ./server 0.0.0.0:50051 0.0.0.0:50052
metrics server listening at 0.0.0.0:50052
gRPC server listening at 0.0.0.0:50051
```

And then run our client:

```
$ go run ./client 0.0.0.0:50051
--------ADD--------
rpc error: code = Unavailable desc = unexpected error:
  unavailable
```

We get one error. While this looks like it only did one query, if you look back at your server, you should be able to see the following:

```
INFO :started call todo.v2.TodoService AddTask
WARN :finished call todo.v2.TodoService AddTask
INFO :started call todo.v2.TodoService AddTask
WARN :finished call todo.v2.TodoService AddTask
INFO :started call todo.v2.TodoService AddTask
WARN :finished call todo.v2.TodoService AddTask
```

That is effectively three requests that were made.

Bazel

As always, you will need to run `gazelle-update-repos` and `gazelle` in order to get the new dependencies and link them to your library:

```
$ bazel run //:gazelle-update-repos
$ bazel run //:gazelle
```

And now you should be able to run your client correctly:

```
$ bazel run //client:client 0.0.0.0:50051
--------ADD--------
rpc error: code = Unavailable desc = unexpected error:
  unavailable
```

To conclude, we saw in this section that we can retry depending on some conditions, with exponential backoff, and for a certain amount of time. Retry is an important feature since the network is often unreliable and we do not want to make the user retry manually each time there is a problem.

Summary

In this chapter, we looked at the key features that we can get by using community projects such as `protoc-gen-validate` or `go-grpc-middleware`. We saw that we can encode request validation logic in our proto files. This makes our code less bloated and provides error message consistency across all the endpoints of our API.

Then, we looked at what middleware are and how to create one. We started with refactoring our authentication and logging interceptors. We saw that by using `go-grpc-middleware`, we can focus only on the actual logic of the interceptor and have less boilerplate to deal with.

After that, we saw that we can expose tracing data from our API. We used OpenTelemetry and Prometheus to gather the data from the gRPC API and expose it through an HTTP server.

We then learned how to apply rate limiting on our APIs. This is helpful to prevent fraudulent actors or defective clients from overloading our server. We used the Token Bucket algorithm and an already existing implementation of a rate limiter to apply limiting to our API.

And finally, we also saw that we can use interceptors on the client side by working with retry middleware. This lets us retry a call depending on an error code, with a maximum number of retries, and optionally with exponential backoff.

In the next chapter, we will go over the development lifecycle for gRPC APIs, how we can ensure their correctness, how we can debug them, and how we can deploy them.

Quiz

1. What is the purpose of the protoc-gen-validate plugin?

    A. Providing checking logic in a `.proto` file

    B. Generating validation code

    C. Both of them

2. What is `go-grpc-middleware` used for?

    A. Providing commonly used interceptors
    B. Generating validation code

3. Which middleware is used for displaying events as human-readable text?

    A. `tracing`
    B. `auth`
    C. `logging`

4. Which middleware is used to constrain the number of requests made per second?

    A. `tracing`
    B. `ratelimit`
    C. `auth`

Answers

1. C
2. A
3. C
4. B

Challenges

- Simplify the client logger you created in the last chapter by using the logging middleware.
- Check the `protoc-gen-validate` rules (`https://github.com/bufbuild/protoc-gen-validate/blob/main/README.md`) and simplify the error handling that you added in the last chapter's challenges.
- Check the other middleware available in `https://github.com/grpc-ecosystem/go-grpc-middleware/tree/v2` and try to implement one. An example could be the selector middleware.
- Create a simple Grafana dashboard based on the metrics the server is exposing. An example could be a dashboard that displays the percentage of requests that succeeded.

9

Production-Grade APIs

Up until now, we've focused on the features provided by gRPC and those added by community projects. That was an important topic but it wasn't the whole story. We now need to think about how to test, debug, and deploy our gRPC server.

In this chapter, we are going to see how to unit and load test our services. Then, we are going to see how we can manually interact with our API to debug it. Finally, we are going to see how we can containerize and deploy our services. This chapter is divided into the following main topics:

- Testing APIs
- Debugging using server reflection
- Deploying gRPC services on Kubernetes

Technical requirements

You can find the code for this chapter in the folder called chapter5 in the companion repo for this book at `https://github.com/PacktPublishing/gRPC-Go-for-Professionals/tree/main/chapter9`. In this chapter, I will be using three main tools: `ghz`, `grpcurl`, and Wireshark. You should already have Wireshark installed from *Chapter 1*, but if this is not the case, you can find it at `https://www.wireshark.org/`. `ghz` is a tool that will let us load test our API. You can get it by visiting `https://ghz.sh/`. Finally, we will use grpcurl to interact with our API from the terminal. You should be able to get it from `https://github.com/fullstorydev/grpcurl`.

Testing

Developing production-grade APIs begins with writing comprehensive tests to ensure that the business requirements are met while also verifying the API's consistency and performance. The first part is mostly handled in unit and integration tests and the second part with load testing.

In the first part of this section, we are going to focus on unit testing the server. We are going to do one test per API type to understand how you can introduce more in the future. In the second part, we are going to introduce ghz, which is a tool for load testing gRPC APIs. We are going to introduce the different options that the tool has and how to load test an API with credentials, an auth token as a header, and so on.

Unit testing

As mentioned, we are going to focus on unit testing the server. Before beginning, it is important to know that the tests presented here are not all the possible tests that we could do. To keep this book readable, I will be presenting how to write unit tests for each API type, and you can find an example of other tests in the `server/impl_test.go` file.

Before writing any tests, we need to do some setup. We are going to write some boilerplate for the different tests to share the same server and connection. This is mostly to avoid creating new servers and connections each time we are running a test. However, note that these are non-hermetic tests. This means that an unexpected state could be shared across multiple tests and make the tests flaky. We are going to introduce ways to deal with this and make sure we clear the states.

The first thing that we can do is create a fake database. This is like what we did with `inMemoryDb`, and in fact, `FakeDb` is a wrapper around `inMemoryDb`, but we are also going to test problems due to connectivity with the database.

To do so, we are going to use the same pattern as `grpc.ServerOption`. `grpc.ServerOption` is a function applying a value to a private struct. An example of this is `grpc.Creds`:

```
func Creds(c credentials.TransportCredentials) ServerOption {
  return newFuncServerOption(func(o *serverOptions) {
    o.creds = c
  })
}
```

It returns a function that, once called, will set the value of c to the `creds` property in `serverOptions`. Note that `serverOptions` is different from `ServerOption`. This is a private struct.

We are going to create a function that tells us whether the database is available or not. Later, we are going to enable the option to return an error if it is not. In `test_options.go`, we will have the following:

```
func IsAvailable(a bool) TestOption {
  return newFuncTestOption(func(o *testOptions) {
    o.isAvailable = a
  })
}
```

I'll leave it up to you to check the rest of the content of `test_options.go`. The functions and structs there simply create some utilities and variables in order to be able to write the `IsAvailable` function and get default values for `isAvailable`.

Now, we can create FakeDb. As mentioned, this is a wrapper around inMemoryDb, and it has some options. In fake_db.go, we can have the following:

```
type FakeDb struct {
  d *inMemoryDb
  opts testOptions
}
func NewFakeDb(opt ...TestOption) *FakeDb {
  opts := defaultTestOptions
  for _, o := range opt {
    o.apply(&opts)
  }

  return &FakeDb{
    d: &inMemoryDb{},
    opts: opts,
  }
}
func (db *FakeDb) Reset() {
  db.opts = defaultTestOptions
  db.d = &inMemoryDb{}
}
```

We can now create a FakeDb in multiple ways:

```
NewFakeDb()
NewFakeDb(IsAvailable(false))
```

We also override the inMemoryDb functions so that our FakeDb implements the db interface and so that we can instantiate a server with this database. Each function of FakeDb follows the same pattern. We check whether the database is available or not; if it is not, we return an error, and if it is, we return the result of inMemoryDb. An example of this is addTask (in fake_db.go):

```
func (db *FakeDb) addTask(description string, dueDate
  time.Time) (uint64, error) {
  if !db.opts.isAvailable {
    return 0, fmt.Errorf(
      "couldn't access the database",
    )
  }
  return db.d.addTask(description, dueDate)
}
```

Now that we have that, we can move one step closer to writing an actual unit test. We now need to create a server. However, we do not want this server to actually use ports on our computer. Using an actual port could make our tests flaky because if the port is already in use, the test would directly return an error saying that it could not create the instance of the server.

To solve that, gRPC has a package called `bufconn` (`grpc/test/bufconn`). It lets us create a buffered connection and thus does not need to use ports. `bufconn.Listen` will create a listener and we will be able to use this listener to server requests. In `server_test.go`, we will share the listener and database as global variables. This will let us dispose of the listener after all tests and add/clear tasks in the database from within a test. On top of that, we will create a function that returns a `net.Conn` connection so that we can use it within the test to create a client:

```go
import (
  "context"
  "log"
  "net"

  pb "github.com/PacktPublishing/gRPC-Go-for-Professionals/
    proto/todo/v2"

  "google.golang.org/grpc"
  "google.golang.org/grpc/test/bufconn"
)

const bufSize = 1024 * 1024

var lis *bufconn.Listener
var fakeDb *FakeDb = NewFakeDb()

func init() {
  lis = bufconn.Listen(bufSize)
  s := grpc.NewServer()

  var testServer *server = &server{
    d: fakeDb,
  }

  pb.RegisterTodoServiceServer(s, testServer)

  go func() {
    if err := s.Serve(lis); err != nil && err.Error() !=
        "closed" {
      log.Fatalf("Server exited with error: %v\n", err)
    }
  }()
}
```

```
func bufDialer(context.Context, string) (net.Conn, error) {
  return lis.Dial()
}
```

The first thing to notice is that we are using the Go init() function to do this setup before the tests are started. Then, notice that we create an instance of our server and register the implementation of our TodoService. Finally, the server is serving in a goroutine. So, we need to make sure that the goroutine is canceled.

We are almost done with the boilerplate. We need to create a client that uses the bufDialer function to connect to the server through the buffered connection. In impl_test.go, we are going to create a function that returns TodoServiceClient and grpc.ClientConn. The first is obviously to call our endpoints but the second one is for us to close the client connection at the end of each test:

```
func newClient(t *testing.T) (*grpc.ClientConn,
  pb.TodoServiceClient) {
  ctx := context.Background()
  creds := grpc.WithTransportCredentials
    (insecure.NewCredentials())
  conn, err := grpc.DialContext(ctx, "bufnet",
    grpc.WithContextDialer(bufDialer), creds)

  if err != nil {
    t.Fatalf("failed to dial bufnet: %v", err)
  }
  return conn, pb.NewTodoServiceClient(conn)
}
```

One important thing to understand here is that we are not testing the whole server that we wrote in main.go. We are simply testing our endpoints implementation. This is why we can connect to the server with insecure credentials. The interceptors, encryption, and so on should be tested in integration tests.

Finally, we can create a small utility function that checks that an error is a grpc error and that it has an expected message:

```
func errorIs(err error, code codes.Code, msg string) bool {
  if err != nil {
    if s, ok := status.FromError(err); ok {
      if code == s.Code() && s.Message() == msg {
        return true
      }
    }
  }
  return false
}
```

We are now ready to write some unit tests. We are going to create a function that will run all the unit tests and dispose of the listener when all the subtests are finished:

```
func TestRunAll(t *testing.T) {
}
```

We are now able to populate the `TestRunAll` function with subtests, like so:

```
func TestRunAll(t *testing.T) {
  t.Run("AddTaskTests", func(t *testing.T) {
    //...
  })

  t.Cleanup(func() {
    lis.Close()
  })
}
```

Let us now write the `testAddTaskEmptyDescription` function, which checks that we get an error when we send a request with an empty description. We will create a new instance of a client, create an empty request, send it to `AddTask`, and finally, check that our error has an unknown code (returned by `protoc-gen-validate`) and that the message is `invalid AddTaskRequest.Description: value length must be at least 1 runes` (also from `protoc-gen-validate`):

```
const (
  errorInvalidDescription = "invalid AddTaskRequest
    .Description: value length must be at least 1 runes"
)

func testAddTaskEmptyDescription(t *testing.T) {
  conn, c := newClient(t)
  defer conn.Close()

  req := &pb.AddTaskRequest{}

  _, err := c.AddTask(context.TODO()), req)

  if !errorIs(err, codes.Unknown, errorInvalidDescription) {
    t.Errorf(
      "expected Unknown with message \"%s\", got %v",
      errorInvalidDescription, err,
    )
  }
}
```

We can then add it to our `TestRunAll` function, like so:

```
func TestRunAll(t *testing.T) {
  t.Run("AddTaskTests", func(t *testing.T) {
    t.Run("TestAddTaskEmptyDescription",
      testAddTaskEmptyDescription)
  }
  //...
}
```

To run this test, we can run the following command in the root folder:

```
$ go test -run ^TestRunAll$ ./server
ok
```

Now, before moving on to looking at how to test streams, let us see how we can test with an unavailable database. This is almost the same as what we did in `testAddTaskEmptyDescription`, but we are going to override the database. Finally, we are going to check that we get an internal error and reset the database (to clear the options):

```
const (
  //...
  errorNoDatabaseAccess = "unexpected error: couldn't
    access the database"
)

func testAddTaskUnavailableDb(t *testing.T) {
  conn, c := newClient(t)
  defer conn.Close()

  newDb := NewFakeDb(IsAvailable(false))
  *fakeDb = *newDb

  req := &pb.AddTaskRequest{
    Description: "test",
    DueDate: timestamppb.New(time.Now().Add(5 *
        time.Hour)),
  }

  _, err := c.AddTask(context.TODO(), req)
  fakeDb.Reset()

  if !errorIs(err, codes.Internal, errorNoDatabaseAccess) {
    t.Errorf("expected Internal, got %v", err)
  }
}
```

We can see that it is easy to test a database failure. That is all for unary RPC. I will let you add testAddTaskUnavailableDb to TestRunAll and look at the other tests for AddTasks in impl_test.go.

We are now going to test ListTasks. We will add some tasks to our fake database, call ListTasks, make sure that there is no error, and check that ListTasks iterated through all the tasks:

```go
func testListTasks(t *testing.T) {
  conn, c := newClient(t)
  defer conn.Close()

  fakeDb.d.tasks = []*pb.Task{
    {}, {}, {}, // 3 empty tasks
  }
  expectedRead := len(fakeDb.d.tasks)

  req := &pb.ListTasksRequest{}
  count := 0

  res, err := c.ListTasks(context.TODO(), req)

  if err != nil {
    t.Errorf("unexpected error: %v", err)
  }

  for {
    _, err := res.Recv()

    if err == io.EOF {
      break
    }

    if err != nil {
      t.Errorf("error while reading stream: %v", err)
    }

    count++
  }

  if count != expectedRead {
    t.Errorf(
      "expected reading %d tasks, read %d",
      expectedRead, count,
    )
```

```
      }
  }
```

There is nothing new in terms of calling the API. We already know all of this from when we wrote the client. However, the main difference here, for this test, is we do not look at the values; we simply assert the time we looped. Of course, you could create more sophisticated tests out of this, but I wanted to show you a simple test on a server streaming API so that you can build upon it.

Next, let us test the client streaming API endpoint. As we are working with the `UpdateTasks` endpoint, we will need to set data in our database. After that, we will basically create an array of `UpdateTasksRequest` in order to change all the items in the database, send the requests, and check that all the updates ran without error:

```
func testUpdateTasks(t *testing.T) {
  conn, c := newClient(t)
  defer conn.Close()
  fakeDb.d.tasks = []*pb.Task{
    {Id: 0, Description: "test1"},
    {Id: 1, Description: "test2"},
    {Id: 2, Description: "test3"},
  }

  requests := []*pb.UpdateTasksRequest{
    {Id: 0}, {Id: 1}, {Id: 2},
  }
  expectedUpdates := len(requests)

  stream, err := c.UpdateTasks(context.TODO())
  count := 0

  if err != nil {
    t.Errorf("unexpected error: %v", err)
  }

  for _, req := range requests {
    if err := stream.Send(req); err != nil {
      t.Fatal(err)
    }

    count++
  }

  _, err = stream.CloseAndRecv()
```

```
   if err != nil {
     t.Errorf("unexpected error: %v", err)
   }

   if count != expectedUpdates {
     t.Errorf(
       "expected updating %d tasks, updated %d",
       expectedUpdates, count,
     )
   }
 }
```

This is similar to the previous test. We used a counter to check that all updates were "applied." In an integration test, you would have to check that the value actually changed in the database; however, because we are in unit tests and we have an in-memory database, checking the actual values would not mean much.

Finally, we will test the bidirectional streaming API. This is a little bit more complex in the testing context, but we are going to tackle the problem step by step. Previously, in the client, when an error happened in a goroutine, we simply ran log.Fatalf to exit. However, here, because we want to keep track of errors and we cannot call t.Fatalf in a different goroutine from the one of the tests, we are going to use a channel of struct called countAndError. As its name suggests, this is a structure containing a counter and an optional error:

```
type countAndError struct {
  count int
  err error
}
```

This is useful because now, we will be able to wait for the goroutine to finish and get a result of the channel. First, let us create the function that sends all the requests. This function is called sendRequestsOverStream and it will be called in a separate goroutine:

```
func sendRequestsOverStream(stream
  pb.TodoService_DeleteTasksClient, requests
    []*pb.DeleteTasksRequest, waitc chan countAndError) {
  for _, req := range requests {
    if err := stream.Send(req); err != nil {
      waitc <- countAndError{err: err}
      close(waitc)
      return
    }
  }
}
```

```
  if err := stream.CloseSend(); err != nil {
    waitc <- countAndError{err: err}
    close(waitc)
  }
}
```

If an error occurs, we will close the waiting channel with an error set in the `countAndError` structure.

Then, we can create the function that reads responses. This function is called `readResponsesOverStream` and will also be called in a separate goroutine:

```
func readResponsesOverStream(stream
  pb.TodoService_DeleteTasksClient, waitc chan
    countAndError) {
  count := 0

  for {
    _, err := stream.Recv()

    if err == io.EOF {
      break
    }

    if err != nil {
      waitc <- countAndError{err: err}
      close(waitc)
      return
    }

    count++
  }

  waitc <- countAndError{count: count}
  close(waitc)
}
```

This time, if everything goes well, the channel will get a `countAndError` with a count set. This count is the same as what we did in previous tests. It checks the number of responses that were collected without error.

Now that we have these two functions, we are ready to write the actual test for our bidirectional streaming API. This is similar to what we did for `ListTasks` and `UpdateTasks`; however, this time, we launch two goroutines, wait for the result, and check that we have no error and a count equal to the number of requests:

```go
func testDeleteTasks(t *testing.T) {
  conn, c := newClient(t)
  defer conn.Close()

  fakeDb.d.tasks = []*pb.Task{
    {Id: 1}, {Id: 2}, {Id: 3},
  }
  expectedRead := len(fakeDb.d.tasks)

  waitc := make(chan countAndError)
  requests := []*pb.DeleteTasksRequest{
    {Id: 1}, {Id: 2}, {Id: 3},
  }

  stream, err := c.DeleteTasks(context.TODO())

  if err != nil {
    t.Errorf("unexpected error: %v", err)
  }

  go sendRequestsOverStream(stream, requests, waitc)
  go readResponsesOverStream(stream, waitc)

  countAndError := <-waitc

  if countAndError.err != nil {
    t.Errorf("expected error: %v", countAndError.err)
  }
  if countAndError.count != expectedRead {
    t.Errorf(
      "expected reading %d responses, read %d",
      expectedRead, countAndError.count,
    )
  }
}
```

With that, we have finally finished testing all the different types of gRPC APIs. Once again, there are more tests that can be done, and other examples are available in impl_test.go. I strongly encourage you to take a look there so you can get more ideas.

After adding all these tests to TestRunAll, you should be able to run them like so:

```
$ go test -run ^TestRunAll$ ./server
ok
```

If you want a more detailed output of what test ran, you can add the `-v` option. This will return something like the following:

```
$ go test -run ^TestRunAll$ -v ./server
--- PASS: TestRunAll
    --- PASS: TestRunAll/AddTaskTests
        --- PASS: TestRunAll/AddTaskTests/
        TestAddTaskUnavailableDb
        --- PASS:
//...
PASS
```

Bazel

In order to run tests with Bazel, you can run Gazelle to generate the `//server:server_test` target:

```
$ bazel run //:gazelle
```

You will then have the target available in `server/BUILD.bazel`, and you should be able to run the following:

```
$ bazel run //server:server_test
PASS
```

If you want to get a more verbose output for your tests, you can use the `-test_arg` option and set it to `-test.v`. It will return something like the following:

```
$ bazel run //server:server_test --test_arg=-test.v
--- PASS: TestRunAll
    --- PASS: TestRunAll/AddTaskTests
        --- PASS: TestRunAll/AddTaskTests/
        TestAddTaskUnavailableDb
        --- PASS:
//...
PASS
```

To conclude, we saw how to test unary, server streaming, client streaming, and bidirectional streaming APIs. We saw that we do not need to use a port on the machine running the test when using `bufconn`. This makes our tests less reliant on the environment it runs on. Finally, we also saw that we can use fakes in order to test our system dependencies. This is a bit out of the scope of this book, but it was important to me to show you that you can write normal tests even if you are using gRPC.

Load testing

Another important step when testing your services is to make sure that they are efficient and can handle a specific load. For this, we use load-testing tools that will concurrently send requests to our service. ghz is a tool that does just that. In this section, we are going to see how to use the tool and some options that we need to set in order to test our API.

ghz is a tool that is highly configurable. Run the following command to see and understand the output:

```
$ ghz --help
```

Obviously, we are not going to use all these options but we will examine the most common ones and the ones that we need to use in some specific cases. Let us start by trying to make a simple call.

> **Important note**
>
> In order to run the following load test, you will need to deactivate the rate-limiting middleware in the `server/main.go` file. You can do so by commenting `ratelimit.`
> `UnaryServerInterceptor` and `ratelimit.StreamServerInterceptor`.

We first run our server:

```
$ go run ./server 0.0.0.0:50051 0.0.0.0:50052
metrics server listening at 0.0.0.0:50052
gRPC server listening at 0.0.0.0:50051
```

The first four options that we are going to talk about are the most common ones. We need to be able to name the service and method that we want to call (`--call`), indicate in which proto file the service is defined (`--proto`) and where to find the imports (`--import_paths`), and finally, specify the data to be sent as a request. In our case, a basic command, run from the `chapter9` folder, will look like this:

```
$ ghz --proto ./proto/todo/v2/todo.proto \
      --import-paths=proto                \
      --call todo.v2.TodoService.AddTask \
      --data '{"description":"task"}'    \
      0.0.0.0:50051
```

However, if you try to run this command, you will end up having an error message like the following:

```
connection error: desc = "transport: authentication
handshake failed: tls: failed to verify certificate: x509:
"test-server1" certificate is not standards compliant"
```

As you can surely guess from the message, this is because we set up our server to only accept secure connections. To solve this problem, we will use the `--cacert` option, which lets us specify a path to where the CA certificate is. If you remember, this is exactly what we did in the code for our client. `ghz` also needs that information:

```
$ ghz #... \
     --cacert ./certs/ca_cert.pem \
     0.0.0.0:50051
```

If you run this command, you will get the same error as previously. This is because a certificate has a domain name associated with it. This means that only requests from a certain domain name will be accepted. However, because we are working from localhost, this simply does not meet that requirement and fails. To solve that, we are going to use the `--cname` option to override the domain name from which we are sending to comply with the certificate:

```
$ ghz #... \
     --cacert ./certs/ca_cert.pem \
     --cname "check.test.example.com" \
     0.0.0.0:50051
```

Here, we used `check.test.example.com` because the generated certificate that we downloaded from `https://github.com/grpc/grpc-go/tree/master/examples/data/x509` was generated with the DNS name `*.test.example.com` (see `openssl.cnf`). Also, note that this `--cacert` and `--cname` are only useful for self-signed certificates. In general, except for specific cases, these certificates are used for testing and non-production environments.

Now, if you run the previous command, you should get the following error:

```
Unauthenticated desc = failed to get auth_token
```

That should ring a bell. This is the error we are sending in our auth interceptor when a client does not provide `auth_token` metadata. In order to send that metadata, we are going to use the `--metadata` option, which takes a JSON string for keys and values:

```
ghz #... \
    --metadata '{"auth_token":"authd"}' \
    0.0.0.0:50051
```

After running with all these options, we should be able to run our first load test (the results might be different for you):

```
$ ghz --proto ./proto/todo/v2/todo.proto \
      --import-paths=proto                \
      --call todo.v2.TodoService.AddTask  \
      --data '{"description":"task"}'      \
```

```
        --cacert ./certs/ca_cert.pem          \
      --cname "check.test.example.com"     \
      --metadata '{"auth_token":"authd"}' \
      0.0.0.0:50051

Summary:
  Count:      200
  Total:      22.89 ms
  Slowest:    16.70 ms
  Fastest:    0.20 ms
  Average:    4.60 ms
  Requests/sec:    8736.44

Response time histogram:
  0.204   [1]    |
  1.854   [111]  |■■■■■■■■■■■■■■■■■■■■■■■■■■■■■■■■■■■■■■■■■■■■■■■■■■
  3.504   [38]   |■■■■■■■■■■■■■■■■
  5.153   [0]    |
  6.803   [0]    |
  8.453   [0]    |
  10.103  [0]    |
  11.753  [0]    |
  13.403  [2]    |■
  15.053  [26]   |■■■■■■■■■■
  16.703  [22]   |■■■■■■■■■

Latency distribution:
  10 % in 0.33 ms
  25 % in 0.78 ms
  50 % in 1.75 ms
  75 % in 2.39 ms
  90 % in 15.12 ms
  95 % in 15.31 ms
  99 % in 16.48 ms

Status code distribution:
  [OK]      200 responses
```

There is a lot to say and to look at in this summary. However, let us focus on some interesting points. The first one is the number of requests made. We can see that we made 200 of them in this test. This is the default number of requests. We can change that by using the - -total option and setting another number (e.g., 500).

Then, in the response time histogram, we can see that 111 out of 200 requests were executed in ~2.29 ms. Another interesting thing to see here is that we have some commands (50) running in more than 13 ms. If we were in production, we might want to dig deeper into this in order to find the cause of these "high" execution times. This depends a lot on the use case and requirements. In our case, this is almost certainly due to the inefficient "database" that we use, or more precisely, the append that we repeatedly call in inMemoryDb.addTask.

After that, we have the distribution of our execution time. We can see that 75% of our requests execute in under 2.39 ms. In fact, this is a similar information as presented previously. If we take the number of requests under 3.504 ms, add them up, and calculate the percentage, we get (1 + 111 + 38) * 100 / 200 = 75%.

Then, we have the status code distribution. In our case, all 200 requests succeeded. However, in a production scenario, you might have something that looks more like this (from the ghz documentation):

```
Status code distribution:
  [Unavailable] 3 responses
  [PermissionDenied] 3 responses
  [OK] 186 responses
  [Internal] 8 responses
```

Finally, one last thing that we cannot see here (because we do not have any) is the error distribution. This is the distribution of the error messages. Once again, in production, you might have something like the following (from the ghz documentation):

```
Error distribution:
 [8]  rpc error: code = Internal desc = Internal error.
 [3]  rpc error: code = PermissionDenied desc = Permission
   denied.
 [3]  rpc error: code = Unavailable desc = Service unavailable.
```

There is obviously a lot more that we could do with this tool. As mentioned, it is highly configurable, and it is even possible to link the results in Grafana (https://ghz.sh/docs/extras) for visualization. However, this is out of the scope of this book. I will leave it up to you to try the different options and call ghz on our other API endpoints to see how they perform.

To conclude, we saw how can load test our service with ghz. We only saw how to use it for our unary API, but it is also useful for testing all the other streaming APIs. After executing the ghz command, we saw that we can get information about latency, error codes, error message distribution, and the fastest and slowest running times. All of this is useful, but it is important to understand that it can be even more powerful when linked with visualization tools such as Grafana.

Debugging

No matter how well we unit test our services, we are humans and humans make mistakes. At some point, we are going to need to debug a service. In this section, we are going to see how to approach debugging. We are first going to enable server reflection, which will let us call our service simply from the command line. After that, we will use Wireshark to inspect data on the wire. Finally, because the error might not always come directly from our code, we will see how we can take a look at the gRPC logs.

Server reflection

Server reflection is an interesting feature when it comes to exposing the API to external clients. This is because it lets the server describe itself. In other words, the server knows all the services registered and the message definition. If a client asks for more information, the server, through reflection, can list all the services, messages, and so on. With that, the client does not even need to have a copy of the proto file. Now, this is not only useful for exposing the API to external clients. It is also useful for manual testing/debugging. It lets developers/testers focus only on debugging the API and not getting the whole environment to work (copy the proto files, and so on).

Enabling server reflection is an easy thing in gRPC Go. We only need two lines of code: an `import` statement and a call to the `reflection.Register` function to register the reflection service on our server. It looks like this (`server/main.go`):

```go
import (
  //...
  "google.golang.org/grpc/reflection"
)

func newGrpcServer(lis net.Listener, srvMetrics
  *grpcprom.ServerMetrics) (*grpc.Server, error) {
  //...
  s := grpc.NewServer(opts...)

  pb.RegisterTodoServiceServer(/*…*/)
  reflection.Register(s)

  return s, nil
}
```

However, even though this is enough to expose information, we will need to get a client that contacts the server and understand the information it is getting. There are multiple such tools out there. The most popular one is grpcurl (`https://github.com/fullstorydev/grpcurl`). If you are familiar with cURL, this is basically a similar tool, but one that understands the gRPC protocol. Even though we are going to use this tool to explore server reflection, know that it can also make other normal requests. If you are interested in such a tool, the repository's README is full of examples of how to use it for other tasks.

Let us try to create a simple command with `grpcurl` first. We are going to use options that are similar to the ones we used in ghz. We are going to use the CA certificate and override the domain name with `-cacert` and `-authority`. Then, we are going to add an `auth_token` header for reflection with `-reflect-header`, and finally, we will use the list verb in order to list the services present on the server:

```
$ grpcurl  -cacert ./certs/ca_cert.pem \
           -authority "check.test.example.com" \
           -reflect-header 'auth_token: authd' \
           0.0.0.0:50051 list
```

Once we run this command, we should get the following output:

```
grpc.reflection.v1alpha.ServerReflection
todo.v2.TodoService
```

We can see that we have both our `TodoService` and the `ServerReflection` service that we registered earlier. With that, we can describe a service to get all the RPC endpoints it contains. We do that with the `describe` verb followed by the service name:

```
$ grpcurl  -cacert ./certs/ca_cert.pem \
           -authority "check.test.example.com" \
           -reflect-header 'auth_token: authd' \
           0.0.0.0:50051 describe todo.v2.TodoService
```

Running this command will show the definition of the service:

```
todo.v2.TodoService is a service:
service TodoService {
  rpc AddTask ( .todo.v2.AddTaskRequest ) returns (
    .todo.v2.AddTaskResponse );
  rpc DeleteTasks ( stream .todo.v2.DeleteTasksRequest )
    returns ( stream .todo.v2.DeleteTasksResponse );
  rpc ListTasks ( .todo.v2.ListTasksRequest ) returns (
    stream .todo.v2.ListTasksResponse );
  rpc UpdateTasks ( stream .todo.v2.UpdateTasksRequest )
    returns ( .todo.v2.UpdateTasksResponse );
}
```

We can also take a look at the message content by replacing the service's name after `describe` with the name message. An example for `AddTaskRequest` is as follows:

```
$ grpcurl -cacert ./certs/ca_cert.pem \
```

```
                -authority "check.test.example.com" \
                -reflect-header 'auth_token: authd' \
                0.0.0.0:50051 describe todo.v2.AddTaskRequest

todo.v2.AddTaskRequest is a message:
message AddTaskRequest {
  string description = 1 [(.validate.rules) = {
    string:<min_len:1> }];
  .google.protobuf.Timestamp due_date = 2
    [(.validate.rules) = { timestamp:<gt_now:true> }];
}
```

Now, as we are talking about debugging in this section, we want to be able to call these RPC endpoints and test them with different data. This is easy because we do not even need to have the proto file with us. The server reflection will help `grpcurl` figure everything out for us. Let us call the `AddTask` endpoint with an invalid request:

```
$ grpcurl -cacert ./certs/ca_cert.pem \
            -authority "check.test.example.com" \
            -rpc-header 'auth_token: authd' \
            -reflect-header 'auth_token: authd' \
            -d '' \
            -use-reflection \
            0.0.0.0:50051 todo.v2.TodoService.AddTask
```

Notice that we use other options here. We use the `-d` option to set the data that we want to send as `AddTaskRequest`. We use the `–use-reflection` option so that `grpcurl` can verify that the data is valid (we are going to see that soon) and the `–rpc-header` on top of `–reflect-header` because `–reflect-header` only sends the header to the `ServerReflection` service, and we also need to send the header to `TodoService`.

As expected, the previous command returns the following error:

```
ERROR:
  Code: Unknown
  Message: invalid AddTaskRequest.Description: value length
    must be at least 1 runes
```

Now, as mentioned, grpcurl does not let us execute commands without adding guardrails. The use of reflection is useful here because it does not let us send data that cannot be deserialized in the request message. An example is as follows:

```
$ grpcurl #... \
            -d '{"notexisting": true}' \
            0.0.0.0:50051 todo.v2.TodoService.AddTask
```

```
Error invoking method "todo.v2.TodoService.AddTask": error
getting request data: message type todo.v2.AddTaskRequest
has no known field named notexisting
```

Finally, as we also have a non-unary RPC endpoint that we would want to test, we can use an interactive terminal. This will let us send and receive multiple messages. To do that, we will set the data to @ and end the command with <<EOF, where EOF stands for end of file (you can use any suffix really). This will let us type data interactively, and when we are finished, we write EOF to let grpcurl know.

Let us start by adding two new tasks to our server:

```
$ grpcurl #... \
          -d '{"description": "a task!"}' \
          0.0.0.0:50051 todo.v2.TodoService.AddTask
$ grpcurl #... \
          -d '{"description": "another task!"}' \
          0.0.0.0:50051 todo.v2.TodoService.AddTask
```

Then, we can use ListTasks to show the tasks:

```
$ grpcurl #... \
              -d '' \
              0.0.0.0:50051 todo.v2.TodoService.ListTasks
{
  "task": {
    "id": "1",
    "description": "a task!",
    "dueDate": "1970-01-01T00:00:00Z"
  },
  "overdue": true
}
{
  "task": {
    "id": "2",
    "description": "another task!",
    "dueDate": "1970-01-01T00:00:00Z"
  },
  "overdue": true
}
```

Can you spot any bugs here? If not, do not worry; we are going to come back to it shortly.

Then, to call our client streaming API (UpdateTasks), we can use the following command on Linux/Mac (press *Enter* after the last EOF):

```
$ grpcurl #... \
          -d @ \
          0.0.0.0:50051 todo.v2.TodoService.UpdateTasks <<EOF
{ "id": 1, "description": "a better task!" }
{ "id": 2, "description": "another better task!" }
EOF
```

Windows (PowerShell) users should use the following:

```
$ $Messages = @"
{ "id": 1, "description": "new description" }
{ "id": 2, "description": "new description" }
"@
$ grpcurl #... \
          -d $Messages \
          0.0.0.0:50051 todo.v2.TodoService.UpdateTasks
```

After that, another call to `ListTasks` should show the data with the new descriptions.

Now, while executing these functions, you might have noticed a bug. If you did not, there is nothing to worry about; we are going to solve the problem together. The problem is that we can send an empty `DueDate`, which then gets transformed to the 0 value in Unix Time (`1970-01-01`). This bug comes from the fact that `protoc-gen-validate` checks `DueDate` only if it is set.

To solve that, we can add another validation rule for `due_date` to our `todo.proto` file. This rule is `required`. It will make it impossible for the field to not be set. You might argue that a task does not need a due date, but we could make a different endpoint for `Adding Notes` and say that `Tasks` should have a due date, but not `Note`.

`due_date` will now be defined as such:

```
google.protobuf.Timestamp due_date = 2 [
  (validate.rules).timestamp.gt_now = true,
  (validate.rules).timestamp.required = true
];
```

We can rerun the generation of the `validate` plugin (in `chapter9`):

```
$ protoc -Iproto --validate_out="lang=go,
paths=source_relative:proto" proto/todo/v2/*.proto
```

Then, we shut our server down and relaunch it:

```
$ go run ./server 0.0.0.0:50051 0.0.0.0:50052
metrics server listening at 0.0.0.0:50052
gRPC server listening at 0.0.0.0:50051
```

If we rerun one of the previous `AddTask` commands, it should fail:

```
$ grpcurl #... \
          -d '{"description": "a task!"}' \
          0.0.0.0:50051 todo.v2.TodoService.AddTask
ERROR:
  Code: Unknown
  Message: invalid AddTaskRequest.DueDate: value is
    required
```

We solved a bug! How great is that?

If you want to now send a request with a `due_data` value you will have to specify a date in `RFC3339` format as a string. An example using a `due_date` value of 500 years from the day of writing this is as follows:

```
$ ghz #... \
    -d '{"description":"task", "due_date": "2523-06-01T14
    :18:25+00:00"}' \
    0.0.0.0:50051
```

Bazel

In order to run the server with Bazel, you will have to update the dependencies. You can run Gazelle to update `//server:server`:

```
$ bazel run //:gazelle
```

Then, you will be able to run the server normally:

```
$ bazel run //server:server 0.0.0.0:50051 0.0.0.0:50052
```

To conclude, we saw that we can turn server reflection on in order to get information and interact with a server for debugging purposes. We saw that we can list services and describe both services and messages. We also saw that we can call unary RPC endpoints, and finally, we saw that we can also call streaming APIs with interactive terminals.

Using Wireshark

Sometimes, we need to be able to inspect the data that is going through the wire. This lets us get a sense of how heavy the payloads are, know if we execute too many requests, and so on. In this section, we are going to see how we can use Wireshark to analyze payloads and requests.

The first thing that we need in order to get access to readable information is to disable the encryption via TLS that we enabled in `Chapter 7`. Note that this should be fine because we are in development mode, but you will need to make sure that encryption is on when you push back to production.

To disable the encryption, we are going to create a switch variable. With this variable, the TLS will be disabled by setting the ENABLE_TLS environment variable to false. Obviously, as we want to make TLS the default, we are going to check whether the environment variable value is different from false, so that if there is a typo in the value or the value is not set, TLS will be enabled.

In server/main.go, we can have the following:

```go
func newGrpcServer(lis net.Listener, srvMetrics
  *grpcprom.ServerMetrics) (*grpc.Server, error) {
  var credsOpt grpc.ServerOption
  enableTls := os.Getenv("ENABLE_TLS") != "false"

  if enableTls {
    creds, err := credentials.NewServerTLSFromFile(
      "./certs/server_cert.pem", "./certs/server_key.pem")
    if err != nil {
      return nil, err
    }
    credsOpt = grpc.Creds(creds)
  }
  //...
  opts := []grpc.ServerOption{/*...*/}
  if credsOpt != nil {
    opts = append(opts, credsOpt)
  }
  //...
}
```

We now need to do something similar on the client side (client/main.go):

```go
func main() {
  //...
  var credsOpt grpc.DialOption
  enableTls := os.Getenv("ENABLE_TLS") != "false"

  if enableTls {
    creds, err := credentials.NewClientTLSFromFile
      ("./certs/ca_cert.pem", "x.test.example.com")
    if err != nil {
      log.Fatalf("failed to load credentials: %v", err)
    }
```

```
  credsOpt = grpc.WithTransportCredentials(creds)
} else {
  credsOpt = grpc.WithTransportCredentials
    (insecure.NewCredentials())
}
//...
opts := []grpc.DialOption{
  credsOpt,
  //...
}
//...
}
```

With that, we can now enable/disable TLS easily. To run the server on Linux or Mac and without TLS, we can now run the following:

```
$ ENABLE_TLS=false go run ./server 0.0.0.0:50051
0.0.0.0:50052
```

For Windows (PowerShell), we can run the following:

```
$ $env:ENABLE_TLS='false'; go run ./server 0.0.0.0:50051
0.0.0.0:50052; $env:ENABLE_TLS=$null
```

Similarly, for the client, we can run the following (Linux/Mac):

```
$ ENABLE_TLS=false go run ./client 0.0.0.0:50051
```

For Windows (PowerShell), we can run the following:

```
$ $env:ENABLE_TLS='false'; go run ./client 0.0.0.0:50051;
$env:ENABLE_TLS=$null
```

We are now ready to start inspecting the data sent over the wire. In Wireshark, we will first check the network interface on which we want to intercept the payloads:

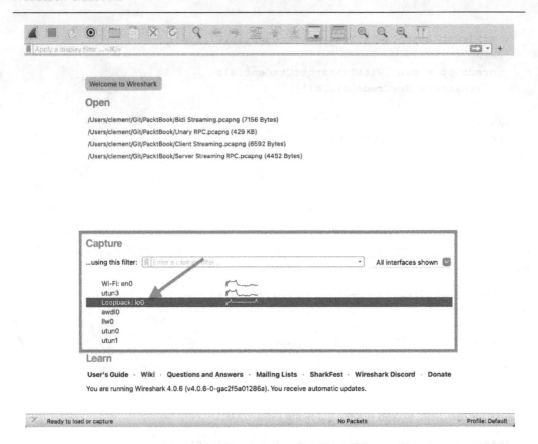

Figure 9.1 – Selecting the network interface

The loopback interface is the one that we are working on: localhost. By double-clicking on it, we will enter the recording interface. But before doing that, we want to tell Wireshark where to find our proto files. Without them, it will show you only the field tags and the value. It would be better if we could have the field names too.

To do that, we will go to **Settings** (or **Preferences** for Mac) | **Protocols** | **Protobuf**, and select a path for the Protobuf search. In our case, we will use the chapter9/proto folder and the folder needed to access the Well-Known Types. The last path depends on how you installed protoc. Here are the most common paths:

- If installed through GitHub releases and you moved the include folder to /usr/local, then the second path is /usr/local/include.

- If installed through `brew`, you should be able to get the path on which protobuf is installed with the `brew --prefix protobuf` command. This will give you a path; simply append `/include` to the path.

- If installed through Chocolatey, you should run the `choco list --local-only --exact protoc --trace` command. This will list a path finishing with `.files`. Open the path in a tool such as Notepad, find a path containing `include/google/protobuf`, and select it up until the `include` folder – for example, `C:\ProgramData\chocolatey\lib\protoc\tools\include`.

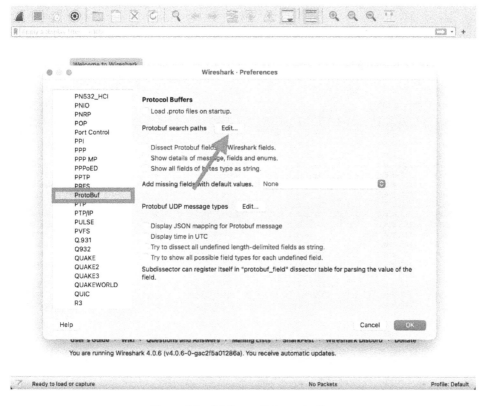

Figure 9.2 – Adding path to proto files

Once this is done, we can go back to our loopback interface and double-click on it. We should now have the following recording interface.

We will then enter a filter to only show the requests on port `50051` and only the requests related to gRPC and Protobuf. Make sure you click on the arrow just next to the filter area; otherwise, you will get all the requests made on the interface.

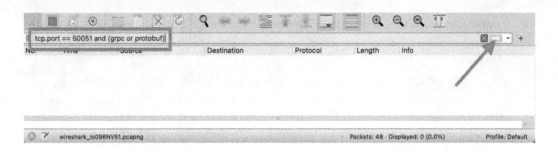

Figure 9.3 – Entering a filter

After that, we can go ahead and run the server and the client. Once the client is done executing, you will have some logs appearing in Wireshark. This should look like the following:

No.	Time	Source	Destination	Protocol	Length	Info
89	30.443062	127.0.0.1	127.0.0.1	GRPC	100	DATA[1] (GRPC) (PROTOBUF)
95	30.443797	127.0.0.1	127.0.0.1	GRPCHTTP2	128	HEADERS[1]: 200 OK, DATA[1] (GRP
100	30.443993	127.0.0.1	127.0.0.1	GRPC	123	HEADERS[3]: POST /todo.v2.TodoSe
107	30.444201	127.0.0.1	127.0.0.1	GRPCHTTP2	94	HEADERS[3]: 200 OK, DATA[3] (GRP
109	30.444280	127.0.0.1	127.0.0.1	GRPC	153	WINDOW_UPDATE[0], PING[0], HEADE
113	30.444399	127.0.0.1	127.0.0.1	GRPCHTTP2	94	HEADERS[5]: 200 OK, DATA[5] (GRP
119	30.444569	127.0.0.1	127.0.0.1	GRPC	110	HEADERS[7]: POST /todo.v2.TodoSe
125	30.446038	127.0.0.1	127.0.0.1	GRPC	115	HEADERS[7]: 200 OK, DATA[7] (GRP
131	30.447130	127.0.0.1	127.0.0.1	GRPC	110	DATA[7] (GRPC) (PROTOBUF)
137	30.448304	127.0.0.1	127.0.0.1	GRPCHTTP2	121	DATA[7] (GRPC) (PROTOBUF), HEADE
142	30.448474	127.0.0.1	127.0.0.1	GRPCGRPCGRPC...	198	HEADERS[9]: POST /todo.v2.TodoSe
149	30.448674	127.0.0.1	127.0.0.1	GRPCHTTP2	92	HEADERS[9]: 200 OK, DATA[9] (GRP
152	30.448773	127.0.0.1	127.0.0.1	GRPC	87	HEADERS[11]: POST /todo.v2.TodoS
160	30.450025	127.0.0.1	127.0.0.1	GRPC	117	HEADERS[11]: 200 OK, DATA[11] (G
166	30.451138	127.0.0.1	127.0.0.1	GRPC	88	DATA[11] (GRPC) (PROTOBUF)
172	30.452332	127.0.0.1	127.0.0.1	GRPC	78	DATA[11] (GRPC) (PROTOBUF)
178	30.452513	127.0.0.1	127.0.0.1	GRPCGRPCGRPC...	113	DATA[13] (GRPC) (PROTOBUF), DATA
186	30.452707	127.0.0.1	127.0.0.1	GRPCGRPCGRPC...	120	HEADERS[13]: 200 OK, DATA[13] (G
192	30.452916	127.0.0.1	127.0.0.1	GRPC	87	HEADERS[15]: POST /todo.v2.TodoS

> Frame 89: 100 bytes on wire (800 bits), 100 bytes captured (800 bits) on interface lo0, id 0

Figure 9.4 – Logs appearing in Wireshark

We can now understand what was sent over the network. If we are looking at the payloads, will should be looking at the DATA (GRPC) (PROTOBUF) frames. An example is the DATA frame for AddTask:

```
Protocol Buffers: /todo.v2.TodoService/AddTask,request
    Message: todo.v2.AddTaskRequest
        Field(1): description = This is another task
          (string)
```

```
Field(2): due_date = 2023-06-01T17:06:20
    .531406+0800 (message)
        Message: google.protobuf.Timestamp
            Field(1): seconds = 1685610380 (int64)
            Field(2): nanos = 531406000 (int32)
            [Message Value: 2023-06-01T17:06:
                20.531406+0800]
```

Finally, if we are looking at gRPC-related recordings, we can take a look at the HEADERS and DATA (GRPC) frame. These can tell you when half-closes and trailers are sent and their size. An example of a half-close for ListTasks is as follows:

```
HyperText Transfer Protocol 2
    Stream: DATA, Stream ID: 7, Length 45
        Length: 45
        Type: DATA (0)
        Flags: 0x00
            0000 .00. = Unused: 0x00
            .... 0... = Padded: False
            .... ...0 = End Stream: False
        0... .... .... .... .... .... ... = Reserved: 0x0
        .000 0000 0000 0000 0000 0000 0000 0111 = Stream
            Identifier: 7
        [Pad Length: 0]
        DATA payload (45 bytes)
```

An example trailer for DeleteTasks is as follows:

```
HyperText Transfer Protocol 2
    Stream: HEADERS, Stream ID: 13, Length 2
        Length: 2
        Type: HEADERS (1)
        Flags: 0x05, End Headers, End Stream
            00.0 ..0. = Unused: 0x00
            ..0. .... = Priority: False
            .... 0... = Padded: False
            .... .1.. = End Headers: True
            .... ...1 = End Stream: True
        0... .... .... .... .... .. .... = Reserved: 0x0
        .000 0000 0000 0000 0000 0000 0000 1101 = Stream
            Identifier: 13
        [Pad Length: 0]
        Header Block Fragment: bfbe
        [Header Length: 40]
```

```
[Header Count: 2]
Header: grpc-status: 0
Header: grpc-message:
```

For the sake of keeping this book readable, we will have to end this section here. However, there is a lot more to look at and discover. We saw that we can use Wireshark to intercept messages over the wire. We made a switch variable to be able to disable TLS temporarily to not read encrypted data. We loaded protobuf messages into Wireshark to let it know how to deserialize messages. Finally, we saw that we can look at messages, as well as lower-level parts of the HTTP2 protocol.

Turning gRPC logs on

Finally, if you are ready to go to an even lower level than Wireshark to debug gRPC applications, gRPC provides two important environment variables to get logs from the framework.

The first environment variable is GRPC_GO_LOG_SEVERITY_LEVEL. It will give you the logs written by gRPC depending on certain severity levels (debug, info, or error). To enable this, you can simply execute your binary or Go command with GRPC_GO_LOG_SEVERITY_LEVEL set in front of it. We did something similar with our custom ENABLE_TLS variable.

An example of running GRPC_GO_LOG_SEVERITY_LEVEL set with info while spinning up the server and closing it is as follows (for Linux/Mac):

```
$ GRPC_GO_LOG_SEVERITY_LEVEL=info go run ./server
  0.0.0.0:50051 0.0.0.0:50052
INFO: [core] [Server #1] Server created
metrics server listening at 0.0.0.0:50052
gRPC server listening at 0.0.0.0:50051
INFO: [core] [Server #1 ListenSocket #2] ListenSocket
created
shutting down servers, please wait...
INFO: [core] [Server #1 ListenSocket #2] ListenSocket
deleted
gRPC server shutdown
metrics server shutdown
```

For Windows (PowerShell), we have the following:

```
$ $env:GRPC_GO_LOG_SEVERITY_LEVEL='info'; go run ./server
  0.0.0.0:50051 0.0.0.0:50052;
  $env:GRPC_GO_LOG_SEVERITY_LEVEL=$null
```

On top of the severity level, you can also set the verbosity of those logs with GRPC_GO_LOG_VERBOSITY_LEVEL, which takes a number between 2 and 99, where the bigger the number, the more verbose it will be. This will not be in a short-term runtime like the one we have right now.

This will be more useful on long-term runs, which we normally have for servers. To enable it, we add GRPC_GO_LOG_VERBOSITY_LEVEL just after our GRPC_GO_LOG_SEVERITY_LEVEL:

```
$ GRPC_GO_LOG_SEVERITY_LEVEL=info GRPC_GO_LOG
  _VERBOSITY_LEVEL=99 go run ./server 0.0.0.0:50051
    0.0.0.0:50052
```

Finally, I know that I said there are two important environment variables but there is another that deserves a mention. This one is important if you are planning to parse logs. You can set the formatter for the logs. As of right now, we have the following:

```
INFO: [core] [Server #1] Server created
```

But we can set the formatter to JSON in order to get the following:

```
{"message":"[core] [Server #1] Server created\n",
  "severity":"INFO"}
```

You will now be able to deserialize the JSON and implement all the kinds of tools you need to monitor and report errors.

To conclude, in this brief section, we saw that we can get information for code that we do not write: the gRPC framework. I am aware that the examples presented in this section are superficial, but generally, these flags are set when something goes really wrong or if you are involved in the development of gRPC Go itself. I still think it is important to know about their existence and I encourage you to try getting more interesting messages out of it.

There are as many ways to debug as there are requirements and settings. As such, we cannot cover everything here, but at least you have the basic skills and tools to get started with hacking. In this section, we saw that we can enable server reflection to get information from the server and interact with it with grpcurl. We also saw that we can intercept messages with Wireshark to get a sense of the requests made and their size. Finally, we saw that we can turn on a certain flag to get logs from gRPC. Before going on to the next section, I wanted to mention that there is another tool that you might find useful and that we did not cover here. This tool is called Channelz (https://grpc. io/blog/a-short-introduction-to-channelz/). Its purpose is to debug networking issues. You might want to take a look at it.

Deploying

Another crucial step of production-grade APIs is deploying the services online. In this section, we will see how we can create a Docker image for gRPC Go, deploy it to Kubernetes, and finally deploy the Envoy proxy to let clients make requests from outside the cluster to a server inside it.

Docker

This first step in deploying is often containerizing your application with Docker. If we did not do so, we would have to deal with errors depending on the server architecture, tools not being available on it, and so on. By containerizing our application, we can build our image once and run it everywhere where Docker is available.

We are going to focus on containerizing our server. This makes much more sense than working on the client because we will later deploy our gRPC server as microservices in Kubernetes and we will make the client, which is outside, make requests to them.

The first thing that we can think about is what all the steps needed to build our application are. We ran it quite a few times, but we need to remember all the tools that we set up in the first place. This includes the following:

- protoc to compile our proto files
- Proto Go, gRPC, and the `validate` plugin to generate Go code out of proto files
- Obviously, Golang

Let us start with getting protoc. For that, we are going to create a first stage based on Alpine, which will use `wget` to get the protoc ZIP file and unzip it inside `/usr/local`. If you are impatient, you can find the whole Dockerfile in `server/Dockerfile`, but we are going to explain it step by step:

```
FROM --platform=$BUILDPLATFORM alpine as protoc
ARG BUILDPLATFORM TARGETOS TARGETARCH

RUN export PROTOC_VERSION=23.0 \
    && export PROTOC_ARCH=$(uname -m | sed
      s/aarch64/aarch_64/) \
    && export PROTOC_OS=$(echo $TARGETOS | sed
      s/darwin/linux/) \
    && export PROTOC_ZIP=protoc-$PROTOC_VERSION-$PROTOC_OS-
      $PROTOC_ARCH.zip \
    && echo "downloading: " https://github.com/
protocolbuffers/protobuf/releases/download/
v$PROTOC_VERSION/$PROTOC_ZIP \
    && wget https://github.com/protocolbuffers/protobuf/
releases/download/v$PROTOC_VERSION/$PROTOC_ZIP \
    && unzip -o $PROTOC_ZIP -d /usr/local bin/protoc
'include/*' \
    && rm -f $PROTOC_ZIP
```

There are quite a few things happening here. First notice that we are using the Docker BuildKit engine. This lets us use defined variables such as BUILDPLATFORM, TARGETOS, and TARGETARCH. We do that because even though we are containerizing our application to avoid dealing with architecture, running a container with the same architecture as the host (virtualization) is much more efficient than emulation. Furthermore, as you can see, we need to specify the architecture and OS in the URL to download protoc.

Then, we define some variables that are important for building up the download URL. We set the version of protoc (here, 23.0). Then, we set the architecture we want to work on. This is based on the result of uname −m, which gives information about the machine. Notice that we use a little trick to replace aarch64 with aarch_64. This is because if you take a look at the releases on the Protobuf repository (https://github.com/protocolbuffers/protobuf/releases), they use aarch_64 in their ZIP filenames.

After that, we use the TARGETOS variable to define which OS we want to deal with. Notice, once again, the similar trick to replace darwin with linux. This is simply because protoc does not have a binary specific to macOS. You can simply use a Linux one.

Then, we do the actual downloading of the file by concatenating all the variables that we defined previously, and we unzip the file into /usr/local. Notice that we are extracting both the protoc binary (/bin/protoc) and the /include folder because the first one is the compiler that we are going to use and the second one is all the files needed to include Well-Known Types.

Now that this is done, we can create another stage for building the application with Go. Here, we are going to copy protoc from the previous stage, download the protoc plugins, compile the proto files, and compile the Go project. We are going to use an Alpine-based image for that:

```
FROM --platform=$BUILDPLATFORM golang:1.20-alpine as build
ARG BUILDPLATFORM TARGETOS TARGETARCH

COPY --from=protoc /usr/local/bin/protoc /usr/local/
bin/protoc
COPY --from=protoc /usr/local/include/google /usr/local/
include/google

RUN go install google.golang.org/protobuf/cmd/protoc-gen-
go@latest
RUN go install google.golang.org/grpc/cmd/protoc-gen-go-
grpc@latest
RUN go install github.com/envoyproxy/protoc-gen-
validate@latest

WORKDIR /go/src/proto
COPY ./proto .

RUN protoc -I. \
    --go_out=. \
```

```
    --go_opt=paths=source_relative \
    --go-grpc_out=. \
    --go-grpc_opt=paths=source_relative \
    --validate_out="lang=go,paths=source_relative:." \
    **/*.proto

WORKDIR /go/src/server
COPY ./server .

RUN go mod download
RUN CGO_ENABLED=0 GOOS=$TARGETOS GOARCH=$TARGETARCH go
  build -ldflags="-s -w" -o /go/bin/server
```

At this point, none of this should be confusing. This is exactly what we have done earlier in the book. However, I want to mention some non-trivial things that are happening here. We are once again taking BuildKit-defined parameters. This lets us use the GOOS and GOARCH environment variables to build a Go binary for this specific setting.

Also, notice that we are copying both protoc and the include folder. As mentioned, the second one is the directory containing Well-Known Types and we use some of them in our proto files, so this is necessary.

Finally, I am using two linker flags. The -s flag is here to disable the generation of the Go symbol table. While I will not dive into what this means, this is sometimes used when creating smaller binaries to remove some information that should not impact the runtime capabilities. -w removes debug information. As these are not needed for production, we can just get rid of them.

Finally, we will build our last stage, which will be based on a scratch image. This is an image that does not have any OS and that we use for hosting binaries and making our images really small. In there, we will copy our certificates into a certs directory, copy the binary we created with go build, and launch the application with the parameters that we usually set:

```
FROM scratch

COPY ./certs/server_cert.pem ./certs/server_cert.pem
COPY ./certs/server_key.pem ./certs/server_key.pem

COPY --from=build /go/bin/server /
EXPOSE 50051 50052
CMD ["/server", "0.0.0.0:50051", "0.0.0.0:50052"]
```

With that, we are ready to build our first image of the server. The first thing that we can create is a Docker Builder. As described in the Docker documentation: "Builder instances are isolated environments where builds can be invoked." This is basically an environment that we need to launch the build of our images. To create that, we can run the following command:

> **Important note**
>
> You need to make sure that Docker is running. This is as simple as making sure that Docker Desktop is running. Finally, you might need to prepend all the following Docker commands with `sudo` if you are on Linux/Mac and you did not create a Docker group and add your user to it.

```
$ docker buildx create --name mybuild --driver=docker-
  container
```

Notice that we give this build environment the name `mybuild` and that we use the `docker-container` driver. This driver will let us generate multi-platform images. We are going to see that later.

Once we have executed the command, we will be able to use this Builder in another Docker command: `docker buildx build`. With this command, we are going to generate the image. We will give it a tag (a name), specify where to find the Dockerfile, specify the architecture we want to build on, and load the image into Docker. To build an image for `arm64` (you can try `amd64`), we run the following (from `chapter9`):

```
$ docker buildx build \
  --tag clementjean/grpc-go-packt-book:server \
  --file server/Dockerfile \
  --platform linux/arm64 \
  --builder mybuild \
  --load .
```

After everything is built, we should be able to see the image by executing the following command:

```
$ docker image ls
REPOSITORY                        TAG      SIZE
clementjean/grpc-go-packt-book    server   10.9MB
```

Finally, let us try to run the server image and make requests to it. We are going to run the image we just created and expose the ports we used for the server (`50051` and `50052`) to the same ports on the host:

```
$ docker run -p 50051:50051 -p 50052:50052
  clementjean/grpc-go-packt-book:server
metrics server listening at 0.0.0.0:50052
gRPC server listening at 0.0.0.0:50051
```

Now, if we run our client normally, we should be able to get all the logs we had previously:

```
$ go run ./client 0.0.0.0:50051
```

To conclude, we saw that we can create slim images around our gRPC applications. We used a multi-stage Dockerfile in which we first downloaded protoc and the Protobuf Well-Known Types. We then downloaded all the Golang dependencies and built a binary, and finally, we copied the binary into a scratch image to create a thin wrapper around it.

Kubernetes

Now that we have our server image, we can deploy multiple instances of our service and we will have created our to-do microservice. In this section, we are going to focus mostly on how to deploy our gRPC service. This means that we are going to write a Kubernetes configuration. If you are not familiar with Kubernetes, there is nothing to be afraid of. Our configuration is simple and I will explain all the blocks.

The first thing that we need to think about is how our service will be accessed. We have two major ways of exposing our services: making them accessible only from inside the cluster or accessible from outside the cluster. In most cases, we do not want our services to be accessed directly. We want to go through a proxy that will redirect and load balance the requests to multiple instances of our service.

As such, we will create a Kubernetes Service, which will itself assign a DNS A record for all the instances of our service. This basically means that each of our services will have its own internal address in the cluster. This will let our proxy resolve all the addresses and load balance across all of them.

Such a service is called a headless service. In Kubernetes, this is a service with the `clusterIp` property set to None. Here is the service definition (`k8s/server.yaml`):

```
apiVersion: v1
kind: Service
metadata:
  name: todo-server
spec:
  clusterIP: None
  ports:
  - name: grpc
    port: 50051
  selector:
    app: todo-server
```

Note that we create a port called `grpc` with the value `50051`. This is because we want to be able to access all the services on port `50051`. Then, notice that we are creating a selector to specify which app this service will handle. In our case, we call it `todo-server` and this will be the name for our deployments.

Now, we can think about creating instances of our service. We are going to do that with a Kubernetes Deployment. This will let us specify how many instances we want, which image to use, and which container port to use. This looks like the following (`k8s/server.yaml`):

```
apiVersion: apps/v1
kind: Deployment
metadata:
  name: todo-server
```

```
    labels:
      app: todo-server
  spec:
    replicas: 3
    selector:
      matchLabels:
        app: todo-server
    template:
      metadata:
        labels:
          app: todo-server
      spec:
        containers:
        - name: todo-server
          image: clementjean/grpc-go-packt-book:server
          imagePullPolicy: Always
          ports:
          - name: grpc
            containerPort: 50051
```

Here, we specify that the name of the Pods will match `todo-server`. This makes them be handled by the service. We then specify that we want to use the image that we created earlier. However, notice here that we are setting `imagePullPolicy` to `Always`. This means that each time we create the Pods, they will pull a new image from the image registry. This makes sure that we always get the newest image on the registry. However, note that this might be inefficient if the images do not change often and if you have local copies of the images that are not outdated. I would recommend you check, depending on your Kubernetes environment, what to use as a value of `imagePullPolicy`. Finally, we use port `50051`. There is nothing more to it than specifying on which port our service is exposing the API.

> **Important note**
>
> For the remainder of this chapter, I expect that you already have a Kubernetes cluster. If you have one in the cloud, this is perfect and you can continue. If you do not have one, you can refer to Kind (`https://kind.sigs.k8s.io/`), and once installed, you can create a simple cluster with the configuration provided in `k8s/kind.yaml`. Simply run `kind create cluster --config k8s/kind.yaml`.

With that, we can now deploy our three services. We will run the following command from the `chapter9` folder:

```
$ kubectl apply -f k8s/server.yaml
```

We are going to execute the following command to look at the Pod being created:

```
$ kubectl get pods
NAME                            READY   STATUS
todo-server-7d874bfbdb-2cqjn    1/1     Running
todo-server-7d874bfbdb-gzfch    1/1     Running
todo-server-7d874bfbdb-hkmtp    1/1     Running
```

Now, since we do not have a proxy, we will simply use the `port-forward` command from Kubernetes to access one server and see whether it works. This is purely for testing purposes, and we are going to see later how to hide the services behind a proxy. So, we run the following:

```
$ kubectl port-forward pod/todo-server-7d874bfbdb-2cqjn
50051
Forwarding from 127.0.0.1:50051 -> 50051
Forwarding from [::1]:50051 -> 50051
```

Then, we should be able to use our client normally on `localhost:50051`:

```
$ go run ./client 0.0.0.0:50051
```

To conclude, we saw that we can use a headless service to create a DNS A record for each of the Pods in the Deployment. We then deployed three Pods and saw that we can test whether they are working or not by using the `port-forward` command in `kubectl`.

Envoy proxy

Now that we have our microservices created, we need to add a proxy that will balance the load between all of them. This proxy is Envoy. This is one of the few proxies that can interact with gRPC services. We are going to see how to set up Envoy to redirect traffic to our services, load balance with the round robin algorithm, and enable TLS.

Let us first focus on writing a listener. This is an entity that specifies the address and port on which to listen and defines some filters. These filters, at least in our case, will let us route the requests for `todo.v2.TodoService` to an Envoy cluster. A cluster is the entity that will let us define the actual endpoints and shows us how to load balance. We can first write our listener (`envoy/envoy.yaml`):

```
node:
  id: todo-envoy-proxy
  cluster: grpc_cluster

static_resources:
  listeners:
  - name: listener_grpc
    address:
```

```
    socket_address:
      address: 0.0.0.0
      port_value: 50051
filter_chains:
- filters:
  - name: envoy.filters.network.http_connection_manager
    typed_config:
      "@type": type.googleapis.com/envoy.extensions
      .filters.network.http_connection_manager.
       v3.HttpConnectionManager
    stat_prefix: listener_http
    http_filters:
    - name: envoy.filters.http.router
      typed_config:
        "@type": type.googleapis.com/envoy.extensions
          .filters.http.router.v3.Router
    route_config:
      name: route
      virtual_hosts:
      - name: vh
        domains: ["*"]
        routes:
        - match:
            prefix: /todo.v2.TodoService
            grpc: {}
          route:
            cluster: grpc_cluster
```

The most important things to note are that we defined a route matching all the gRPC requests from any domain names and matching the /todo.v2.TodoService prefix. Then, all these requests will be redirected to grpc_cluster.

After that, let us define our cluster. We are going to use the STRICT_DNS resolution to detect all the gRPC services by DNS A record. Then, we will specify that we are only accepting HTTP/2 requests. This is because, as you know, gRPC is based on HTTP/2. After that, we will set the load balancing policy to use round robin. Finally, we will specify the address and port of the endpoint:

```
clusters:
- name: grpc_cluster
  type: STRICT_DNS
  http2_protocol_options: {}
  lb_policy: round_robin
  load_assignment:
    cluster_name: grpc_cluster
```

```
     endpoints:
   - lb_endpoints:
     - endpoint:
       address:
         socket_address:
           address: "todo-server.default.svc
             .cluster.local"
           port_value: 50051
```

Notice that we use the address generated by Kubernetes. This is of the form `$SERVICE_NAME-$NAMESPACE-svc-cluster.local`.

In order to test our configuration, we can run everything locally first. We will temporarily make the `listener_0` port equal to `50050` so that it does not conflict with our server port:

```
static_resources:
  listeners:
  - name: listener_grpc
    address:
      socket_address:
        address: 0.0.0.0
        port_value: 50050
```

We will also have to set the endpoint address to localhost to access the server running locally:

```
- endpoint:
  address:
    socket_address:
      address: 0.0.0.0
      port_value: 50051
```

Then, we will run our server:

```
$ go run ./server 0.0.0.0:50051 0.0.0.0:50052
metrics server listening at 0.0.0.0:50052
gRPC server listening at 0.0.0.0:50051
```

We can now run our envoy instance with `func-e`:

```
$ func-e run -c envoy/envoy.yaml
```

Finally, we can run our client on port `50050`, not `50051`:

```
$ go run ./client 0.0.0.0:50050
--------ADD--------
2023/06/04 11:36:45 rpc error: code = Unavailable desc =
```

```
last connection error: connection error: desc = "transport:
authentication handshake failed: tls: first record does not
look like a TLS handshake"
```

As you can guess, this is because Envoy is somehow breaking the TLS connection between the server and the client. To solve that, we are going to specify that the upstream of our cluster uses TLS and that the downstream of our listener also uses TLS.

In the filters, we will tell Envoy where to find our self-signed certificates:

```
#...
filter_chains:
- filters:
  #...
  transport_socket:
  name: envoy.transport_sockets.tls
  typed_config:
    "@type": type.googleapis.com/envoy.extensions
      .transport_sockets.tls.v3.DownstreamTlsContext
    common_tls_context:
      tls_certificates:
        - certificate_chain:
            filename: /etc/envoy/certs/server_cert.pem
          private_key:
            filename: /etc/envoy/certs/server_key.pem
```

Note that this is probably not what you would do in production. You would use a tool such as Let's Encrypt to automatically generate your certificates and link them.

Now, we will tell the cluster that the upstream is also using TLS:

```
clusters:
- name: grpc_cluster
  #...
  transport_socket:
    name: envoy.transport_sockets.tls
    typed_config:
      "@type": type.googleapis.com/envoy.extensions
        .transport_sockets.tls.v3.UpstreamTlsContext
```

Obviously, this is not going to work directly. On our local computer, we do not have the /etc/envoy/certs/server_cert.pem and /etc/envoy/certs/server_key.pem files. But we have them in the chapter9 certs folder. We will replace them temporarily:

```
- certificate_chain:
    filename: ./certs/server_cert.pem
```

```
    private_key:
      filename: ./certs/server_key.pem
```

Let us now kill the previous instance of Envoy and rerun it:

```
$ func-e run -c envoy/envoy.yaml
```

Finally, we should be able to run our client and receive responses from our server:

```
$ go run ./client 0.0.0.0:50050
```

We are now certain that our requests go through Envoy and are redirected to our gRPC server. The next step will be reverting all the temporary changes that we made for testing (listener port to 50051, endpoint address to todo-server.default.svc.cluster.local, and certs path to /etc/envoy) and creating a Docker image that we will use for deploying Envoy in our Kubernetes cluster.

To build such an image, we will copy the certificates to /etc/envoy/certs (once again, this is not recommended in production) and the configuration (envoy.yaml) to /etc/envoy. Finally, this image will run the envoy command with the --config-path flag, which will point to the /etc/envoy/envoy.yaml path. In envoy/Dockerfile, we have the following:

```
FROM envoyproxy/envoy-distroless:v1.26-latest
COPY ./envoy/envoy.yaml /etc/envoy/envoy.yaml
COPY ./certs/server_cert.pem /etc/envoy/certs/
server_cert.pem
COPY ./certs/server_key.pem /etc/envoy/certs/server_key.pem
EXPOSE 50051
CMD ["--config-path", "/etc/envoy/envoy.yaml"]
```

We can now build the image for arm64 (you can use amd64) like so:

```
$ docker buildx build \
    --tag clementjean/grpc-go-packt-book:envoy-proxy \
    --file ./envoy/Dockerfile \
    --platform linux/arm64 \
    --builder mybuild \
    --load .
```

That is it! We are ready to deploy Envoy in front of our TODO microservices. We need a headless service for Envoy. This is for the same reasons that we had when creating a headless service for our microservices. In production, there will potentially be more than one instance of Envoy and you need to make sure they are all addressable. In envoy/service.yaml, we have the following:

```
apiVersion: v1
kind: Service
metadata:
```

```
  name: todo-envoy
spec:
  clusterIP: None
  ports:
  - name: grpc
    port: 50051
  selector:
    app: todo-envoy
```

Then we need to create a `Deployment`. This time, as we are in a development setting, we will deploy only one Pod for Envoy. All the rest of the configuration is similar to what we did with our gRPC server. In `envoy/deployment.yaml`, we have the following:

```
apiVersion: apps/v1
kind: Deployment
metadata:
  name: todo-envoy
  labels:
    app: todo-envoy
spec:
  replicas: 1
  selector:
    matchLabels:
      app: todo-envoy
  template:
    metadata:
      labels:
        app: todo-envoy
    spec:
      containers:
      - name: todo-envoy
        image: clementjean/grpc-go-packt-book:envoy-proxy
        imagePullPolicy: Always
        ports:
          - name: grpc
            containerPort: 50051
```

We can now run all of this. I am assuming that you did not tear down the previous step that we did for deploying microservices. Right now, you should have the following:

```
$ kubectl get pods
NAME                           READY   STATUS
todo-server-7d874bfbdb-2cqjn   1/1     Running
todo-server-7d874bfbdb-gzfch   1/1     Running
```

```
todo-server-7d874bfbdb-hkmtp    1/1    Running
```

So, now we can first add the service and then the deployment for Envoy:

```
$ kubectl apply -f envoy/service.yaml
$ kubectl apply -f envoy/deployment.yaml
$ kubectl get pods
NAME                            READY   STATUS
todo-envoy-64db4dcb9c-s2726     1/1     Running
todo-server-7d874bfbdb-2cqjn    1/1     Running
todo-server-7d874bfbdb-gzfch    1/1     Running
todo-server-7d874bfbdb-hkmtp    1/1     Running
```

Finally, before running the client, we can use the `port-forward` command to forward Envoy's port `50051` to `localhost:50051`:

```
$ kubectl port-forward pod/todo-envoy-64db4dcb9c-s2726
  50051
Forwarding from 127.0.0.1:50051 -> 50051
Forwarding from [::1]:50051 -> 50051
```

We can then run the client and we should be able to get some results:

```
$ go run ./client 0.0.0.0:50051
//...
error while receiving: rpc error: code = Internal desc =
unexpected error: task with id 1 not found
```

Notice that because of the load balancing and the fact that we do not use a real database, the Pods are not able to find tasks that are stored in other Pods' memory. This is normal in our case, but in production, you would be relying on a shared database and these problems would not arise.

To conclude, we saw that we can instantiate Envoy in front of our services to redirect requests with a certain load-balancing policy. This time, contrary to the load balancing we saw in chapter 7, the client does not actually know any addresses for the servers. It connects to Envoy and Envoy is redirecting requests and responses. We obviously did not cover all the possible configurations for Envoy and I would recommend that you check out other features, such as rate limiting and authentication.

Summary

In this chapter, we covered unit and load testing. We saw that we can find bugs and performance issues by extensively testing different parts of our system. Then, we saw how to debug our application when we found a bug. We used server reflection and grpcurl to interact with our API from the terminal. Finally, we saw how we can containerize our services and deploy them on Kubernetes. We saw that we can create headless services to expose our microservices with a DNS A record per gRPC server, and we saw that we can put Envoy in front of them to do load balancing, rate limiting, authentication, and so on.

Quiz

1. What tool is useful for load testing?

    A. Wireshark

    B. grpcurl

    C. ghz

2. In Wireshark, what information can you look at?

    A. gRPC HTTP/2 frames

    B. Protobuf messages

    C. All of them

3. What is Envoy used for?

    A. Redirecting requests and responses

    B. Logging

    C. Exposing metrics

    D. Load balancing

    E. A and D

    F. B and C

Answers

1. C

2. C

3. E

Challenges

- Add support for a real database. You should be able to do so by implementing the db interface and creating an instance of your struct in the registered server instance.

- Expose the Prometheus metrics in your Kubernetes cluster. You can take a look at prometheus-operator (https://github.com/prometheus-operator/prometheus-operator).

Epilogue

As we reach the end of this book on building gRPC microservices in Golang, I hope you have found gRPC Go interesting and useful, and that you are willing to try it on your next project. This book is the book I wish I had when I started learning this awesome technology and I hope this helped you in any way.

Throughout this book, we both explored some elements of theory and some practical implementations of gRPC services. From learning the networking concepts to the pure implementation and tool that we can use, passing by learning useful considerations when designing an API, you learned the most important skills that you will need for your career as backend engineer.

To conclude this book, I would like to invite you to stay up to date with all the topics related to gRPC and Protobuf. You can do that by following GitHub Topics, read some blog post or simply getting involved in some open source project. This is a fascinating area of backend engineering that needs more attention, more help by building tools, and more people to form communities all around the world.

Thank you for accompanying me on this journey to make production-grade gRPC APIs. I wish you the very best in your future endeavors. May you create innovative and effective APIs.

Happy engineering!

Index

A

AddTask
  calling, from client 84, 85
  implementing 83
ALTS authentication
  reference link 146
API calls
  logging 168-172
  retrying 182-184
  retrying, with Bazel 184
  tracing 172-178
  tracing, with Bazel 179
APIs
  securing, with rate limiting 179-182

B

Bazel 61, 86, 162
  server, running with 209
  used, for authenticating requests 168
  used, for generating Go code 61-63
  used, for handling errors 128
  used, for retrying API calls 184
  used, for running tests 199
  used, for securing APIs with rate limiter 182
  used, for securing connections 148, 149

  used, for tracing API calls 179
  used, for validating requests 162-164
  using, in client boilerplate 70
  using, in server boilerplate 66, 67
bidirectional streaming 10
bidirectional streaming API 98, 99
  database, evolving 99
  DeleteTasks, calling from client 101-103
  DeleteTasks, implementing 99-101
Buf 59
  used, for generating Go code 60
Buf CLI
  protoc-gen-validate, using with 161
  used, for validating requests 161

C

call
  canceling 129-132
CallOption
  reference link 145
Certificate Authority (CA) 71, 146
client boilerplate 68, 69
client-side load balancing 149
client streaming API 92
  database, evolving 92, 93
  UpdateTasks, calling from client 95-98
  UpdateTasks, implementing 94, 95

client streaming RPC 9
common types 34
  Google common types 35
  well-known types 34
connections
  securing 146-148
  securing, with Bazel 148, 149

D

deadlines 133
  specifying 133, 134
debugging, production-grade APIs 204
  gRPC logs, turning on 216, 217
  server reflection 204-209
  server, running with Bazel 209
  Wireshark, using 209-216
DeleteTasks
  calling, from client 101-103
  implementing 99-101
deployment, production-grade APIs 217
  Envoy proxy, adding 224-230
  with Docker 218-221
  with Kubernetes 222-224
descriptor 44
deserialization 22
DialOptions 70
dialoptions.go 70
Docker
  production-grade APIs,
    deploying on 218-221

E

encoding
  reference link 142
Envoy proxy 224
  adding 224-230

errors
  handling 124-128
  handling, with Bazel 128

F

FieldMasks 115, 116
  adopting, for payload reduction 115
field tags 32, 33
  messages, splitting 111-113
  required/optional 109-111
  selecting 109
  UpdateTasksRequest, improving 113-115

G

Go code
  Bazel, using 61-63
  Buf, using 59, 60
  generating 58
  protoc, using 58, 59
Google common types 35
GraphQL 49
gRPC 40
  client code, generating 46, 47
  importance 49
  server code, generating 44-46
  use cases 40
  versus REST APIs 49
  working 42-44
**gRPC, comparing with GraphQL
    and REST APIs**
  convenience of using technology 51
  data format 50
  data schema 50
  developers' workflow 51
  differentiators 50
  SOC of API endpoints 51
grpc.Creds 71

gRPC-Go
reference link 137
grpc.*Interceptor 71, 72
gRPC Load Balancing
reference link 149
gRPC project
prerequisites 55
setting up 55
grpc.SetTrailer function
reference link 134
gzip Compressor
reference link 142

H

HTTP/2 2, 3
server push 2

I

integer type
selecting 105-108
interceptor 164, 165
external logic, applying with 137-142
Interface Description Language (IDL) 20-22
inter-process communication (IPC) 40

K

Kubernetes
production-grade APIs,
   deploying on 222-224

L

length-delimited types 31
ListTasks
calling, from client 90, 91
implementing 88-90

ListTasksRequest 116
improving 116-118
load balancing 149
used for distributing requests 149-153

M

metadata
sending 134-137
middleware 164, 165

P

packed repeated field 119
payload
compressing 142-146
production-grade APIs
debugging 204
deploying 217
testing 187
Protobuf 19, 20, 41, 42
common types 34
fixed-size numbers 27, 28
integer type, selecting 105-108
scalar types 26
services 35, 36
types 26
Protobuf, versus JSON 23
readability 24, 25
schema strictness 25
serialized data size 23, 24
protoc 58
used, for generating Go code 59
protoc-gen-validate
using, with Buf CLI 161
PROTOC_GEN_VALIDATE_VERSION 162
.proto file definition
creating 56-58

R

rate limiting
used, for securing APIs 179-182
read/write flow 47, 48
requests
authenticating 165-168
authenticating, with Bazel 168
distributing with load balancing 149-153
validating 157-161
validating, with Bazel 162-164
validating, with Buf CLI 161
REST APIs 49
constraints 49
RPC life cycle 11
bird's eye view 12
client side connection 14, 15
connection, creating 13, 14
server side connection 15
RPC operations 3
Send Half Close operation 6
Send Header operation 4
Send Message operation 5
Send Trailer operation 6, 7
RPC types 7
bidirectional streaming 10
client streaming RPC 9
server streaming RPC 8, 9
unary RPC 7

S

scalar types 26
schema-driven development (SDD) 51
Send Half Close operation 6
Send Header operation 4
Send Message operation 5
Send Trailer operation 6, 7

separation of concerns (SOC) 40
serialization 22
server boilerplate 63-66
Bazel, using 66, 67
server.go 70
ServerOption 70
server reflection 204
enabling 204
server streaming API 86, 87
database, evolving 87, 88
ListTasks, calling from client 90, 91
ListTasks, implementing 88-90
server streaming RPC 8, 9
strings
using 108

T

task 76
properties 76
template
using 75, 76
testing, production-grade APIs 187
load testing 200-203
unit testing 188-198
with Bazel 199
type embedding 46

U

Unary API 76-78
AddTask, calling from client 84, 85
AddTask, implementing 83
Bazel 86
code generation 78
generated code, inspecting 78, 79
service, registering 79-82
Unary RPC 7

unpacked repeated field 119, 120

UpdateTasks

   calling, from client 95-97

   implementing 94, 95

UpdateTasksRequest

   improving 113-115

V

Validate() function 158

varints 28, 29

   data distribution 31

   negative numbers, need for 30, 31

   range of numbers 30

   selecting 30

W

well-known types 34

Wireshark

   for debugging production-
      grade APIs 209-216

wire types 32, 33

WithCancel function

   reference link 129

Packtpub.com

Subscribe to our online digital library for full access to over 7,000 books and videos, as well as industry leading tools to help you plan your personal development and advance your career. For more information, please visit our website.

Why subscribe?

- Spend less time learning and more time coding with practical eBooks and Videos from over 4,000 industry professionals

- Improve your learning with Skill Plans built especially for you

- Get a free eBook or video every month

- Fully searchable for easy access to vital information

- Copy and paste, print, and bookmark content

Did you know that Packt offers eBook versions of every book published, with PDF and ePub files available? You can upgrade to the eBook version at Packtpub.com and as a print book customer, you are entitled to a discount on the eBook copy. Get in touch with us at customercare@packtpub.com for more details.

At www.packtpub.com, you can also read a collection of free technical articles, sign up for a range of free newsletters, and receive exclusive discounts and offers on Packt books and eBooks.

Other Books You May Enjoy

If you enjoyed this book, you may be interested in these other books by Packt:

Domain-Driven Design with Golang

Matthew Boyle

ISBN: 9781804613450

- Get to grips with domains and the evolution of Domain-driven design
- Work with stakeholders to manage complex business needs
- Gain a clear understanding of bounded context, services, and value objects
- Get up and running with aggregates, factories, repositories, and services
- Find out how to apply DDD to monolithic applications and microservices
- Discover how to implement DDD patterns on distributed systems
- Understand how Test-driven development and Behavior-driven development can work with DDD

Event-Driven Architecture in Golang

Michael Stack

ISBN: 9781803238012

- Understand different event-driven patterns and best practices
- Plan and design your software architecture with ease
- Track changes and updates effectively using event sourcing
- Test and deploy your sample software application with ease
- Monitor and improve the performance of your software architecture

Packt is searching for authors like you

If you're interested in becoming an author for Packt, please visit `authors.packtpub.com` and apply today. We have worked with thousands of developers and tech professionals, just like you, to help them share their insight with the global tech community. You can make a general application, apply for a specific hot topic that we are recruiting an author for, or submit your own idea.

Share Your Thoughts

Now you've finished *gRPC Go for Professionals*, we'd love to hear your thoughts! Scan the QR code below to go straight to the Amazon review page for this book and share your feedback or leave a review on the site that you purchased it from.

`https://packt.link/r/1837638845`

Your review is important to us and the tech community and will help us make sure we're delivering excellent quality content.

Download a free PDF copy of this book

Thanks for purchasing this book!

Do you like to read on the go but are unable to carry your print books everywhere? Is your eBook purchase not compatible with the device of your choice?

Don't worry, now with every Packt book you get a DRM-free PDF version of that book at no cost.

Read anywhere, any place, on any device. Search, copy, and paste code from your favorite technical books directly into your application.

The perks don't stop there, you can get exclusive access to discounts, newsletters, and great free content in your inbox daily

Follow these simple steps to get the benefits:

1. Scan the QR code or visit the link below

https://packt.link/free-ebook/9781837638840

1. Submit your proof of purchase
2. That's it! We'll send your free PDF and other benefits to your email directly